HERBERT BUTTERFIELD
Writings on Christianity and History

HERBERT BUTTERFIELD

Writings on Christianity

and History

Edited with an Introduction by

C. T. McINTIRE

New York

Oxford University Press

1979

Copyright © 1979 Oxford University Press, Inc.

LIBRARY OF CONGRESS CATALOGING IN PUBLICATION DATA

Butterfield, Herbert, Sir, 1900-
  Herbert Butterfield: writings on Christianity and
history.

  "Bibliography: writings by Herbert Butterfield on
Christianity and history": p.
  Includes index.
    1. History (Theology)—Addresses, essays, lectures.
I. McIntire, C. T.   II. Title: Writings on Chris-
tianity and history.
BR115.H5B825   901   78-16756
ISBN 0-19-502454-0

Essay 1 from *Steps to Christian Understanding*, edited by R. J. W. Bevan
(Oxford University Press, London, 1958), pp. 105-21. Reprinted by permission of the publisher.

Essay 2 from *History* XL (1955), 1-17. Reprinted by permission of The Historical Association.

Essay 3 from *Orbis* X (1966-67), 1233-46. Reprinted by permission of *Orbis*.

Essay 9 from *History and Human Relations*, pp. 131-57. Copyright 1951,
1952 by Herbert Butterfield. Reprinted by permission of Collins Publishers
(London) and Harold Matson Co., Inc. (New York).

Essays 10, 11, 12, 13, from *The Christian Newsletter*, numbers 333 (16 March
1949), 88-96; 336 (27 April 1949), 136-44; 341 (6 July 1949), 215-23, 224-32.
Copyright © 1949, The Christian Frontier Council, London, and reprinted
by permission.

Printed in the United States of America

Editor's Preface

In these seventeen essays Herbert Butterfield, Emeritus Regius Professor of Modern History at Cambridge University, touches on many aspects of the general theme of Christianity and history. The topics popularly associated with the theme are raised here: God's work in history, Jesus and history, the Resurrection, salvation history, the historicity of the biblical accounts, the meaning of history, Christianity in history. These he relates to less often considered questions: the Old and New Testaments as historical writing, the various scriptures' interpretation of ancient historical events, the interpretation of world history both during and after the biblical epoch, the rise of a Christian view of history, and the use of Christian insights in historical study and in the explanation of historical events. He takes the theme further still by disclosing quite candidly his own religious presuppositions as a historian, and by offering an assessment of the present and the future of Christianity.

Butterfield brings to these themes the knowledge and expertise of a historian. The result is a freshness of approach and insight. He contributes to the discussion of theology and history, while he also provides elements for a comprehensive development of a

Christian interpretation of history as well as a Christian approach to historical study.

As the title suggests, these essays may be viewed as a sequel to Butterfield's *Christianity and History*, his widely read and classic book on the subject, published in 1949. Like *Christianity and History*, these essays—or most of them—originated as lectures before a university audience. They were written at various times over nearly a thirty year period, beginning with the earliest in 1949 and the latest in 1977. Nine are previously unpublished, while most of the rest were published in specialized periodicals with limited readerships.

My Introduction discusses Butterfield's work on Christianity and history. The special theme is the character and development of his religion and its connection with his work as a historian and religious thinker. The emphasis is upon understanding the sources, context, and main features of his thought; I have made no attempt to criticize his thought or writings. The Introduction may be read as a case-in-point of how a historian's ideas and beliefs about human nature and human history intrinsically affect the make-up of his or her historical writing.

I suggested this collection to Butterfield in the fall of 1975. I had already studied his published writings for a year in my seminar in philosophy of history and historiography at the Institute for Christian Studies, Toronto. When I met with him in June 1975 at Peterhouse, Cambridge, he very kindly gave me several of his unpublished essays to read. Soon afterward I realized that a volume containing his unpublished essays and previously published articles on Christianity and history would be immensely valuable to have. In May 1977 at Peterhouse he and I talked together for three weeks about his life and work. We tape-recorded almost twenty hours of interviews, and he very generously gave me unrestricted access to his diaries, notes, correspondence, and unpublished manuscripts. Most of the personal information about him in my Introduction is derived from the interviews and his papers. It might be mentioned that other unpublished writings relevant to the theme remain in his papers. The essays which appear in this volume I have selected and arranged with his kind approval. Butterfield has willingly read the introduction and checked the biographical details; but I should stress that I am entirely responsible for what is said there. It is not an "official" statement, or even one with which he might find himself in full agreement.

A number of people freely shared with me their insight and their personal knowledge of Butterfield. I am especially thankful for help from Owen Chadwick, Gordon Rupp, and Maurice Cowling in Cambridge; Denis Mack Smith in Oxford; and Patrick Cosgrave in London. I am grateful as well to Allen Kelley in New York who enthusiastically supported the project from the beginning, and to the Fellows of Peterhouse for their generous hospitality.

Above all I am deeply thankful to Herbert Butterfield for the opportunity to work with him on the project. The experience has been a genuine blessing for me. I have found him to be an inspiring person who tries to live the virtues and beliefs he advocates in these essays.

The hope is that the appearance of these essays will benefit a variety of people, both specialists and general readers. As anyone knows who has read Herbert Butterfield's work before, there is rich insight to be gained here as well as some very pleasurable reading.

Toronto   C. T. McIntire
September 1978

Contents

III   Christianity in the Twentieth Century

Introduction
Herbert Butterfield on Christianity and History
By C. T. McINTIRE

Historian and Religious Thinker

Herbert Butterfield was a shy soul in his younger days, and never more shy than when he spoke of his own spirituality and religious faith. Yet even before he finished high school he found himself signed up as a lay preacher on the Methodist circuit around his home village of Oxenhope in Yorkshire, England. His father and the local parson both thought Herbert would be splendid on the circuit. His faith was settled, his competence unmistakable, and his sense of responsibility immense. In spite of his shyness, he preached three or four times quarterly to the villagers in neighboring hamlets, occasionally to good-sized congregations. He wrote out his sermons word for word. His themes, he recalls, were more often devotional and moral than doctrinal. This was during the final years of the First World War.

For twenty years altogether he preached in this fashion. In 1919 he shifted to the circuit around Cambridge, England, and preached in fenland chapels several times annually, first during three years as an undergraduate at Peterhouse, Cambridge, then as a graduate student and a research fellow at Peterhouse in the 1920's, and later in the 1930's as a young lecturer in history at Cambridge University. He actually went up to Cambridge thinking he might become a Methodist parson, although he also had ambitions as a writer. Since at least Herbert's early teens his father had intimated that he badly wanted him to become a preacher, perhaps fulfilling in some way his father's own unsatisfied desire to be a preacher. The fellows of Peterhouse, impressed as they were with Butterfield's brilliance, slowly stole him away from his ministerial purpose even before he finished his B.A. in 1922, and he found himself becoming a historian. In 1936 he gave up lay preaching, although his name remains on the list of local preachers for the Cambridge circuit to this day. It is important to note that in neither case did his change mean any slackening of faith or any loss of spirituality.

In spite of all his experience and intentions he had not lost his reticence and humility in matters of religion. In 1933 he wrote in his diary that public display and sensationalism in religion were like a lover "who speaks of his mistress with vulgarity," parading something personal tastelessly in the street. He commented,

> I think perhaps a Christian should talk little of his religion lest he fall into this sin. It is the part of a spiritual man to be austere with his thoughts and to know what is fitting. Insensitiveness on this point means a real lack of spiritual apprehension.

His hesitation in making public utterance, which perhaps rested on a characteristic of his psychology, reflected his understanding of the nature of Christianity itself. Butterfield, it seems, has always regarded Christianity as a spiritual thing, residing in the inner personality, related directly to God, and moving outward to suffuse the entirety of a person's life and work. His diary entry in 1933 continued,

> I think that the best evidence of Christianity in the heart is quiet assurance, and a flame that burns in silence, and a charity for ever expressing itself, for ever unexpressed; and with these a serene orderliness and a calm reliance on providence.

Butterfield had the makings of a monk.

We can well understand Butterfield's reticence when the Divinity Faculty of Cambridge University began pressing him soon after the Second World War to give a series of public lectures on Christianity and history. By now he was well established as a historian. In 1944 he had become the Professor of Modern History and helped to restore the Cambridge History Faculty whose activity the war had disrupted. He had six books behind him, including a monograph on diplomatic history, *The Peace Tactics of Napoleon, 1806–08* (1929), and *The Whig Interpretation of History* (1931), a book-length essay in historiography of incredible originality. He was a center of vitality for young history students at Peterhouse, and he was widely admired as a superb lecturer to undergraduates.

Throughout the years he had kept his explicitly religious activity generally quite separate from his professional work. Yet people knew him to be a devout Christian. On Sunday mornings he faithfully attended Wesley Methodist Church on Christ's Pieces, Cambridge; he often worshipped Sunday evenings and during the week at Peterhouse Chapel with its Anglican liturgy; and in other quiet ways he let undergraduates in on his piety. Of course many knew of his years as a lay preacher out in the fens. For more than twenty years—until as late as 1944—he taught ecclesiastical history on the side at Wesley House, the Methodist theological college at Cambridge.

There were perhaps three or four explicit religious references in his historical writings before 1948—passages which clearly suggested what his own religious convictions might be. In the last chapter of *Whig Interpretation* (1931) he urged historians to exercise love and understanding toward historical figures rather than issuing moral judgments against them. In the same book his chief example of the complexity of the historical process was the coming of religious liberty, and he talked matter-of-factly about "providence" in history, although he chose not to capitalize the word. *The Englishman and His History* (1944) contained a final chapter clearly approving the salutory influence of Christianity in the English political tradition. And in 1942 he reviewed William Temple's *Christianity and Social Order* and disagreed with Temple's interpretation of the influence of protestantism on capitalism, putting more emphasis on the relationship between secularization and capitalism. He broadcast a couple of talks on the BBC in 1947 and 1948 that referred to religion. He had, however, never published anything advocating Christian belief or discussing how Christian faith related to the study or interpretation of history.

When the Divinity Faculty first approached Butterfield he offered to help them locate a suitable lecturer on Christianity and history. Someone was needed to fill a Saturday morning spot in which previously had been presented a very popular series by a philosopher on Christianity and philosophy. The only names Butterfield put forward were historians who were also clergy. The divines responded that they needed a non-clergyman, whose field was not church history and whose reputation was that of a historian and not a theologian. He was the one to do it, they insisted. They wanted the series to commence in the fall of 1948, in Michaelmas term.

Butterfield finally accepted the duty himself as a call from Providence. He had only the summer of 1948 to write the lectures, for in the spring of 1948 he had to deliver an extra set of lectures on the history of modern science. A Cambridge committee on the history of science, chaired by Joseph Needham, had already conscripted him into that task. Twice in one year he let other people push him out on a limb into areas he did not consider his own.

Butterfield was no theologian, and he had never dreamed of acting as if he were a religious thinker. He had read very little theology and felt himself uninformed about trends in twentieth-century theology and biblical studies. What he did have, at age forty-eight, was a lifetime of his own Bible reading, deep personal belief, and an extraordinary historian's knowledge of human history. Moreover, he had, in fact, given the theme of Christianity and history some thought over the years.

The lectures were a wild success. For seven Saturday mornings during October and November, 1948, perhaps eight hundred dons and students listened to Butterfield talk about the limits of historical study, human personality, sin and cupidity, judgment, cataclysm, Providence, Christianity, the Promise, and Jesus. The BBC broadcast six of his lectures during the spring of 1949, and in October 1949 Butterfield published the lectures, revised, in the book *Christianity and History*. Not a few of his professional colleagues thought the man had gone off, perhaps like some mad elf in a metaphysical fantasy. He wondered whether he was ruining himself for life.

He accomplished a stupendous feat during 1948 and 1949. In addition to *Christianity and History*, he reworked and published his lectures on science as *The Origins of Modern Science*, and he completed and published a highly technical monograph in eighteenth century studies, *George III, Lord North, and the People, 1779–80*. His publisher, Bell of London, released all three books in October 1949. Butterfield was at the top of his career as a historian. *George III, Lord North, and the People* was well received among specialists, while *Origins of Modern Science* and *Christianity and History* sold widely among general readers and students, and were influential in reshaping many things in the two fields to which they pertained. He was eager to write more history.

As Butterfield so often says, we must look at what happens next from the side of those who did not know the outcome of their forays. What Butterfield had not foreseen was that his reluctant adventure in religious thought inaugurated a new career as a religious thinker. And what he regarded as his very personal statement about religion and history formed part of and contributed to a significant cultural movement of renewed interest in a Christian view of history.

Between 1948 and 1956 Butterfield lectured and wrote extensively on themes ramifying from his *Christianity and History*. He worked out his thought on religion, the nature of history and historiography, the history of historiography, and the state of affairs in the twentieth century. He constructed a larger body of thought on Christianity and history than any other historian, except perhaps Christopher Dawson. Among theologians and philosophers, perhaps only Reinhold Niebuhr outdid him in sheer mass of material on a Christian view of history.

Butterfield was drawn along primarily by invitations to lecture or write a book. The post-war period was a time requiring reconstruction of the whole civilization and the rebuilding of personal lives after the catastrophe of Hitler, the depression, and the war, and amid the terror with which Stalinism appeared to endanger the world. People needed what Butterfield had to say and they pulled it out of him.

An important ecumenical forum, *The Christian Newsletter*, requested something from him in 1949 and received four little essays discussing academic history, the biblical interpretation of history, the ecclesiastical interpretation of history, and the Marxian interpretation of history (see essays 10–13). His Riddell Lectures at Durham University in 1951 were published as *Christianity in European History*. The same year he collected eight essays under the title *History and Human Relations*, including "Christianity and Human Relationships," an essay on love as a dynamic in history—which he later wished he had published as part of *Christianity and History*—and "The Christian and Historical Study" (see essay 9). Also in 1951 came the essay "The Scientific versus the Moralistic Approach in International Affairs," an important statement on freedom and necessity in history. For Queen's University in Canada he presented lectures on religion and individual liberty which he published as *Liberty in the Modern World* (1952). For a thousand high school students at a Church of England conference he lectured on "God in History" in 1952, perhaps the best brief summary of his views on personality, laws, and Providence in history (see essay 1). *Christianity, Diplomacy and War* (1952) began as lectures for a Methodist forum, and presented his views of twentieth-century international history, starting with World War I. "The Role of the Individual in History," a discussion of human personality and the reality of individual choice, was an important address to the Historical Association in 1953 (see essay 2). His fullest statement on Christianity and the study of the history of historical study he gave as the Wiles Lectures in Belfast, published in 1955 as *Man on His Past*. At Bristol University in 1956 he delivered a very personal assessment of Christianity in the twentieth century (see essays 14–16). Indeed, ten of the seventeen pieces in this volume of his essays on Christianity and history derive from the period 1948–1956. Meanwhile, the book that started him on his excursion appeared in an American edition and was translated into German, French, Spanish, Italian, Chinese, and four Scandanavian languages.

At the end of what amounted to an eight-year immersion in themes related to Christianity and history, Butterfield still resisted becoming a religious thinker. He acknowledged in his Bristol lectures in 1956 (see essay 14):

> It was never my intention to set up as a teacher in religion, and I always have the feeling that heaven will strike me dead if I show presumption in this respect. On the other hand it is arguable that the Christian ought not to be unwilling to make confession of what he believes; so that I am between the upper and the nether stone, for I also have the feeling that heaven will strike me dead if by repeated refusals I seem to be declining to testify to the faith.

Uncomfortable though he might be, he seems to have enjoyed the work and thrived on it, and his contribution was considerable. Butterfield may think of himself as a historian and disclaim any expertise outside of history, but he perhaps will not think ill of others if they see him as a historian *and* a religious thinker.

After 1956, while his output of historical writing slowed down, he occasionally worked further on religious themes. We may notice, incidentally, that the Fellows of Peterhouse made him Master of the college in 1955, a post he retained until his retirement in 1968. Between 1963 and 1968 he was also Regius Professor of Modern History. In 1960, based on lectures given in America, he published *International Conflict in the Twentieth Century: A Christian View*, an important statement on morality and laws in history. He delivered the Gifford Lectures in Scotland in 1965 and 1966 on the history of historical writing. He examined the Old Testament and the New Testament as historical writing, the rise of a Christian interpretation of history (see essay 7), the tradition of Christan historiography, and the secularization of historiography. Some of these formed the basis of lectures on Christianity and history elsewhere, including Northwestern University in 1974 (see essays 4, 5, 6, 8). There were numerous other lectures on related themes and occasional articles during the sixties and seventies (see essays 3 and 17). He contributed an amazingly concise, yet perceptive, overview of the history of Christianity to *The Dictionary of the History of Ideas* (1973), entitled "Christianity in History."

Butterfield's writings merged into a larger intellectual movement and spiritual renewal during and after the Second World War. Christianity enjoyed a renaissance among some English intellectuals and writers. A very diverse collection of people were united around the theme that Christian faith offered hope for the revitalization and reorientation of a distraught civilization. One remembers the names of C. S. Lewis, T. S. Eliot, Charles Williams, J. R. R. Tolkien, and Dorothy Sayers. Thinking of historians brings to mind Christopher Dawson, David Knowles, R. H. Tawney, Gordon Rupp, and Arnold Toynbee. In church and social thought were people like John Baillie, Kathleen Bliss, Charles Coulson, V. A. Demant, and Stephen Neill. Turning to Europe and America recalls Reinhold Niebuhr, Paul Tillich, John C. Bennett, Georges Florovsky, Kenneth Scott Latourette, Karl Barth, Emil Brunner, Jacques Ellul, Joseph Hromadka, Hendrikus Berkhof, Jacques Maritain, Teilhard de Chardin, and many others. Included in this renaissance was a renewal of a predominately Augustinian Christian view of history, and Butterfield's work contributed to it significantly.[1]

Nonconformist and New Whig

Butterfield was born and raised an English Nonconformist in religion and has remained one all his life. He is a Methodist, and his Methodism is the key to understanding his Christianity.

In 1900 when Butterfield was born, the West Yorkshire village of Oxenhope, with its population of about two thousand, was very nearly a Methodist village, and it was largely a product of the industrial revolution. A number of textile mills which provided the people's livelihood bordered the stream running through the village. There were three Methodist chapels, two very small Baptist Churches, and a Church of England parish. The Methodist chapel which the Butterfield family attended was newly built in the 1890's and could, on special occasions, accommodate perhaps a thousand souls who would come from around the area. The mill masters were Methodists as were a sizeable number of mill workers. Yorkshire was a center for English Methodism.

Oxenhope was almost surrounded by moors, the same moors that Charlotte and Emily Brontë had loved to write about when they had lived in Haworth about a mile and a half along the road north of Oxenhope. The nearest town was Keighley (pronounced Keithley), a manufacturing center less than four miles north on the edge of the Aire River. Bradford, an industrial city, sprawled to the east about nine or ten miles.

Butterfield considers his father, Albert Butterfield, to be the single most influential person in his life, certainly the most important religious influence. His father, for financial reasons, had been forced to leave school at age ten to work as a wool sorter. He educated himself, grew into a devout Methodist, and was always a conscientious worker. He attracted the attention of John Parker, the mill master, and about the time of Herbert's birth, Parker invited Albert to leave the mill floor and begin work as a clerk and bookkeeper in the office. He had married Ada Mary Buckland who was living in with the Parkers to care for their blind son. He became something of a favorite of the mill master who visited the Butterfield household often, giving them books, music, and magazines and helping them buy an upright piano—things which were important for Herbert's early years. Herbert, the firstborn son, had a younger sister and a brother.

Albert Butterfield, instead of becoming a Methodist minister, had to be content with becoming a Leader of a Wesleyan "class meeting." Throughout Herbert's childhood and youth, his father each week led a class meeting of about twelve to sixteen young mill workers, often bringing them to the house. It was a kind of spiritual cell group which gathered to read the Bible, share spiritual experiences, and develop a strong personal bond among themselves.

The Butterfields attended church each Sunday—two services and two Sunday School classes. Herbert greatly disliked the Sunday School, he recalls, but he did appreciate the worship. He developed there the habit of regular Sunday worship which he has continued all his life.

Probably the most significant religious occasions for Herbert were not the church or class meetings, but his frequent walks with his father. From about age seven or eight until age fourteen he and his father walked together often in the evening. As he remembers it, his father would talk—about nature, his ambitions, God, faith, something they had read. He spoke quietly and stated his views almost like a confession of faith. Herbert loved these walks and was devoted to his father. He compares his father's influence to that of Jesuits over the young in former days. Herbert's mother dominated the household and sometimes seemed to regard his father as rather weak. Herbert, however, saw his father as a very powerful man, and he submitted himself to his father's influence. It might be fair to say that his father became the religious and personal touchstone in his life, and it was chiefly these walks that made him so.

The Christianity that surrounded Butterfield during his childhood and youth in Oxenhope might be described as a liberal version of evangelical Methodism. There was no great emphasis on doctrine, but the church and his father believed in the main teachings of orthodoxy—God and Providence, the redeemer Christ born divine and human, the resurrection of Christ, the reality of sin in the human heart, the need for personal salvation, and the ultimate authority of the Bible. These beliefs his father held with tolerance, and without fear of liberalizing movements. Of greater importance to his father—typical of Methodism—was the cultivation of the inner life and the devotional side of spirituality. He valued what they would call the real religious experience of communion with God. The free will of each person was important. His father encouraged Herbert in the virtues of self-education and hard work, of individual choice and achievement, but always in a spirit of humility and charity toward others. It was a Methodist ethic which found the English liberal political and social tradition congenial.

This discussion of his father's influence brings to mind one of Butterfield's comments on historical causation. He wrote, "Of two products of the same house the one became a nonconformist minister because his father was a nonconformist minister; the other became an enemy of religion because his father was a nonconformist minister" (see essay 2). The point notwithstanding, it does appear that Butterfield grew into what were essentially his father's beliefs and ethic, made them his own by age fourteen or sixteen, and has never fundamentally deviated from them since. To be sure, his beliefs developed considerably after his mid-teens, and other people influenced his religious life, but the vital heart of his spirituality and ethic seems to have been set during his teenage years. He never experienced essential religious doubt, although he rebelled often against the practices and institutionalizations of churches and clergy. The bond he achieved with his father seems to have been an orienting line by which he made his way through the tangles of growing into maturity, of developing through school and university, of entering into his career, and of experiencing the crises in the culture after World War I. He remained close to his father until his father died in 1952. What his father thought mattered to him throughout the years. Butterfield testifies that "nothing ever diminished his influence."

At age eleven Herbert started going each day to the Trade and Grammar School in Keighley; he attended there until he was eighteen. The school stressed science, math, and other studies which led the boys into careers in commerce, engineering, and industry. Herbert worked diligently and usually stayed at the top of his class, but he was not enticed by the specialties of the school.

Since as early as perhaps age eight he aspired to be a writer—a writer of novels and poetry—and later on this desire seemed to coexist with his thoughts about becoming a minister. He was an avid reader in his youth who made his way through the books which the mill master had passed along. He read countless novels—the Brontës, Hardy, Hugo, Dumas, G. A. Henty's historical novels, a whole bevy of more faceless Victorian novels, and the Coral Islands literature which abounded in those days. He worked through the Harmsworth Self-Educator which had come in fortnightly installments, designed for the self-help of workers just like his father. He was always writing. He admits writing, or at least beginning, his own "Victorian" novels, and many other pieces of prose which he never showed to anyone. It seems that in those years he never told his father of his ambition to become a writer. None of this writing of his youth remains.

The burdens of homework after age fourteen and the preparations for examinations (the "O levels") at age sixteen cut into the time otherwise given to his clandestine writing, his extra reading, and the regular walks with his father. He wished to study classics for his advanced level examinations (the "A levels"), chiefly to prepare himself for writing, but the headmaster, who wanted him for mathematics, maneuvered him into history as a compromise. He passed the A levels at age eighteen in history, French, and English. As it turned out, he won a scholarship to Peterhouse in history; he would have sought one in English if there had been one, again to suit him for writing. When he found himself becoming a historian he consoled himself by thinking of it partly as opening the way to a career as a writer.

Butterfield's chief teacher during his last years at Keighley was important for his religious formation, although in a way that was quite unexpected. He experienced F. C. Moore as a very able and stimulating teacher in English and history, but he also found him to be incessantly hostile to Christianity. Herbert resisted this side of his teacher's influence, and in the process strengthened his own religious convictions, and clarified his beliefs and ethic.

When he went up to Peterhouse in 1919 Butterfield stepped into a world very different from Oxenhope and Keighley. At Peterhouse the prevailing presence was socially upper class, Anglican in religion, very sophisticated and self-confident. He came from what was really a working-class background with some lower middle-class tendencies. He was a Nonconformist, even a teetotaler, from a small Yorkshire village. In Peterhouse society it was not difficult for him to feel rather awkward and somewhat unsure of himself.

It appears that what he did was to fasten on to an active personal religious life, on one hand, while he stretched himself incredibly in his academic development on the other. These may have been two different worlds for him, his inner world and his outer world. At this time in his life he made no attempt to think through the relationship of these worlds. It was only much later that he began to connect them to one another in his serious thinking, and he did not work out his thought on the issue until 1948 when he had to prepare his lectures on Christianity and history.

His explicitly religious life—his inner world, so to speak—centered on the practice of a spiritual devotion that involved daily Bible reading, contemplation of the inner life, and regular worship and praise of God—all typical of Methodist piety. He involved himself actively in Wesley Methodist Church and gave himself to regular preaching on the lay circuit around Cambridge. Even after he abandoned his own ministerial prospects he assisted others to become Methodist parsons by his teaching of ecclesiastical history at Wesley House, a task which, by its subject matter, was suitably separate from his research on Napoleon at Peterhouse.

It may be that none of Butterfield's sermons as a lay preacher survive, but his diary, which he kept irregularly, and his notes give us a glimpse of his spirituality and the emotion of his inner life during the 1920's. There we see him extolling religion and poetry as two expressions of the soul:

> . . . it is in religion that life has reached its highest points throughout history—in religion, and perhaps in poetry, where the experience is akin to the religious. (probably May 1922)

He experienced God deeply and richly and very personally, almost as would a visionary:

> I saw God round the corner yesterday. It was where the avenue of trees cuts into Trumpington Road and there was a pretty piece of shrubbery, a momentary play of sunshine, a pause in the wind. Then God came without a sound. He was an elf. Quickly he disappeared—and as long as I looked at the shrub I could not make it look like that again, I could not recapture the first elusive vision. I could not find where God had gone. So I felt utterly alone, and friendless, and away from home. Yet in my heart was a song, for I had seen a fabulous thing. (4 February 1926)

And a week later he gives a hint of the tension that he could experience between his inner and outer worlds:

> I can quite understand the lure of the cloister, the charm of a
> lonely sheltered life, the attraction of a retreat from the
> world. . . . Yet to be in the whirl of it, maddeningly engaged
> in it, and lifted by the floods, is the supreme exhilaration. . . .
> (13 February 1926)

During his undergraduate days he felt swirling around him a religious turmoil raised by the proponents of liberal and rationalist
theology. For about two years he identified with a group of the
more liberal Student Christian Movement (SCM) at Peterhouse,
serving for a time on its committee, but he also sometimes attended
meetings sponsored by the evangelical Cambridge Intercollegiate
Christian Union (CICU). He broke with the SCM in his third
year after he was dissatisfied with a committee discussion of
whether belief in God was necessary for membership in the
group. At Wesley House during the 1920's he found theological
liberals gathered in great strength, and he soon took a position
against them. He insisted that there was a center pole of Christian
belief which should stand above the wind and weather, however
flexible one might be in the rest of the matters of religion. Moreover he believed that the theological liberals were too worldly-
minded and, as he wrote later, "would have tied Christianity to
the things that happened to be fashionable in the year 1900" (see
essay 14).

He broke away from the liberals but he did not fling himself
into the arms of the evangelicals. He continued to admire the
originality and the flexibility of the liberals in biblical criticism
and theological thought, and he regarded the evangelicals as too
constricted in just these things. At Princeton in the United States,
where he spent a year in 1924-25, he was particularly impressed
with the intolerance of the American fundamentalists then bat-
tling the modernists in religion. What he admired in some evangelicals was their desire to hold fast to essentials of Christianity,
and their attention to the well-being of the spiritual life. His appreciation of evangelicals was enhanced in 1923 when he met
Carey Francis, a brilliant scholar in mathematics at Peterhouse and

a "CICUite." The two became close associates in religion during the next four years until Carey left for Africa to teach in an Anglican school for boys. Butterfield regards him as "about the most wonderful man I ever met," and confesses that he was "utterly inspired by him."

What Butterfield sought in the 1920's was a Christianity that was bound to neither a liberal nor a strictly evangelical party. What he found for himself he was able to define in his diary in 1932, and it sounded a lot like what John Wesley might have hoped for in the eighteenth century:

> To live a life of piety is inwardly to trust God and often to have communion with him and also to place one's treasure in heaven. The fruits of this are contentment and reconciliation within the self, and the acquisition of inner life—the building up of a fund of spiritual resources, and the deepening of personality. The blossom is in charity that overflows to all men, and in a life that is lived humbly in the world. In all this there is something very difficult for sophisticated men. And it is utter foolishness to those who are worldly wise. (27 August 1932)

There were three, perhaps four, people who, in his experience, especially exemplified such a saintly Christianity—Carey Francis, his friend in the 1920's; David Knowles, the Catholic historian and monk who became a Fellow of Peterhouse in the 1940's and later preceded him as Regius Professor of Modern History at Cambridge; his father; and a quiet old woman in Oxenhope named Martha Whitaker. It was at the end of the twenties that he married Pamela Crawshaw, the daughter of the Methodist parson at Cottenham, north of Cambridge.

In what might be called his outer world, Butterfield committed himself to his studies at Cambridge as an undergraduate. He describes his transmutation from an aspiring litterateur and ministerial prospect into a historian as a process of being carried along by the overpowering influences of Peterhouse. He hoped to study with Harold Temperley, the great diplomatic historian at Peterhouse, known for his research on George Canning, but Temperley, initially unimpressed with him, sent him to another teacher. Later Butterfield wrote an essay on an assigned theme, "Art Is History Made Organic," which some fellows of the college considered so brilliant that they began talking about him as a future fellow. Temperley became impressed, too, and now took him on as his student. Thus began an apprenticeship which lasted several years. Temperley intimidated the young scholar. Butterfield describes how he would present to his teacher brief essays on assigned themes, but week after week Temperley would merely glance at them and make no comment. Then Temperley would run off on some talk of his own—about politics, some anecdote, a poet—heedless of its irrelevance to the subject of the moment. Butterfield needed support and sympathetic understanding from Paul Vellacott, another Peterhouse historian, to cope with the great man's formidability. Nevertheless, it was Temperley chiefly who made him into a historian. Butterfield decided to become a research student in 1922, then accepted a research fellowship in 1923, and his route into history as a career was set. From Temperley Butterfield learned the historical method as developed in the tradition of Ranke—how to find and work with evidence, the scientific criticism of documents, and a trust in the precision and finality obtainable in the study and writing of diplomatic history. It was during the twenties that he latched onto Ranke and made him one of the two names on which he most depended throughout his career.

Butterfield's romantic, literary side developed as well during these years. Between 1919 and 1923, he recalls, at the end of each day he read novels, plays, poetry, and countless detective novels. He soaked himself in great literature to the lasting benefit of his style, and he composed romantic poetry of his own. "Poetry is passion bubbling over, and the flame is imagination," he wrote in an undergraduate essay. He won the Le Bas Prize in 1923 for an essay on the fortuitously assigned theme "The Historical Novel," which became his first published book (1924). After composing the essay, he remembers, he felt it almost impossible to read another historical novel thereafter.

His literary tendency inclined him toward G. M. Trevelyan's side of the debate with J. B. Bury, a controversy on the nature of history which raged later in the 1920's around Cambridge. Trevelyan promoted history as a species of literature, as "art added to scholarship." He advocated the historians' use of sympathetic imagination to penetrate the inner personalities of by-gone figures, he emphasized the importance of narrative and story in history writing, and he insisted on the necessity of good style.[2] All this Butterfield found congenial, and he enlarged his appreciation of Carlyle. He reworked the text of *The Peace Tactics* in order to improve his style, and was pleased when critics noted the literary quality of the book. He later placed himself partly in the tradition of Trevelyan and Carlyle in an essay on "History as a Branch of Literature," in *History and Human Relations* (1951).

However, Butterfield was somewhat attracted to Bury's side too. Bury, Regius Professor of Modern History from 1902 until 1927—and, curiously, whom Trevelyan succeeded—believed the important thing to stress about history was its features as a science. Bury thought well of the Rankeian tradition of scientific source criticism and microscopic research, but he wanted to take the question of history as a science further. There was, Bury argued, a realm of laws in history, the king of which was the law of cause and effect. He thought that if the data accumulated by research could be arranged according to chains of cause and effect, historians could achieve greater scientific objectivity. Bury became interested in the points at which various events correspondent with laws would intersect—the contingencies of history which he attributed to chance.[3] Butterfield was probably stimulated by Bury's argument to reflect on the laws, conditions, and conjunctures of history.

Butterfield's outlook as a historian developed as, almost intuitively, he applied a reconciling method to the debates and conflicts he encountered. It could even be called a form of dialectic method. When confronted by opposing sides of an issue he endeavored to move above the formulations of the problem as presented by the protagonists. He would, imaginatively, let himself be attracted by each side in order to understand them almost from the inside. Next he could sort out his criticisms of each side's errors and exaggerations, and discern the residue of truth in each position. He could then reconcile the truths at a higher level of insight. This is what he did earlier with the liberal-evangelical conflict, and now he applied the approach to Trevelyan, Bury, and Temperley. Later, in the 1930's, he did a similar thing with Marxism. Out of it came Butterfield's distinctiveness as a historian.

One of Temperley's most durable gifts to Butterfield was to put him on to Lord Acton. Acton, who was Regius Professor of Modern History from 1895 to 1902, was celebrated around Cambridge for introducing the scientific method from the German tradition. Butterfield did not submerge himself in Acton during the twenties—that came much later in the 1940's—but he did feel Acton's unmistakable pull on him. Acton was Rankeian in his later years; he was a virtuoso as a historian and lecturer; he exercised a superintending influence over numerous students of history; he possessed sheer brilliance of insight—no doubt all these features attracted Butterfield to him. But the thing that might especially have carried the appeal to him was that Lord Acton was known as a Christian. Like Ranke, here was a historian whose Christianity mattered for his historical writing and lecturing. He was a Catholic who no doubt appreciated the riches of a deep inner life, who spoke easily of Providence, who possessed catholicity of mind, and who above all valued human liberty and the individual person in history as the embodiment of the soul. But there was a flaw in it. Acton also claimed that the historian had a prophetic function which qualified him to pronounce moral judgments upon historical figures—not merely condemnations of the results of their actions, but judgments against them as people. And this disturbed Butterfield's sensitivities as a Methodist who rated humility and the exercise of charity toward others as among the highest virtues. His understanding of the nature of historical study sharpened as he came to the view that the historian's task was humble and limited, and that moral judgments did not belong to the historian's competence as historian.

Something central to his view of the structure of history fitted into place about this same time as well. For all his appreciation of Trevelyan as a literary historian, he came to criticize his outlook as a "Whig" historian. Partly promoted by a disagreement with Trevelyan's *Lord Grey of the Reformed Bill* (1920) and partly due to his reflection on his own practice, he noted that a common feature of much historical writing was its tendency to see things running in a straight line of progress from somewhere in the past and culminating in whatever the historian favored most about the present. In particular he thought that what was currently called the Whig interpretation of history tended too greatly to credit the Protestant Whigs in past centuries as the fathers of English political and religious liberty and toleration. The problem would be worsened when, like Acton, the historian would issue moral judgments against those who could be regarded as enemies of the line progressing toward liberty.

Butterfield churned all this over in his mind, and, as had long been his habit, wrote out his ideas in his notes. In particular he discussed it with Vellacott. In 1931 he published the little book, *The Whig Interpretation of History*. The final chapter was primarily directed against Acton on moral judgments. The remainder perhaps could not be construed as a polemic against Trevelyan, although Trevelyan certainly thought that it was—"I am the last Whig historian," he exclaimed. Butterfield's own proposal about the process of history took off from Ranke's affirmation of the interconnectedness of everything in history and from Bury's point about contingencies. As Butterfield argued, changes in history emerge out of a curious dialectic of different, clashing, or intersecting wills and circumstances:

> It is not by a line but by a labyrinthine piece of network that one would have to make the diagram of the course by which religious liberty has come down to us, for this liberty comes by devious tricks and is born of strange conjunctures, it represents purposes marred perhaps more than purposes achieved, and it owes more than we can tell to many agencies that had little to do with either religion or liberty.

The book won him a mistaken reputation as an anti-Whig. Butterfield himself favored the Whig political tradition over the Tory—which *The Englishman and His History* (1944) later made amply clear. As early as his undergraduate years he developed what became a lifelong fascination with the personality of Charles James Fox, the great Whig politician. Moreover, he felt close to Trevelyan, the literary Whig, and he deeply honored Lord Acton, who could be called a Catholic Whig of sorts. The Whig tradition and moderate English liberalism comported well with the beliefs and ethic he absorbed from his father's Methodist Nonconformity. One ardent Tory scholar at the time, Charles Smythe, commented to Butterfield with considerable insight that no consolation could be drawn from the book for a *Tory* interpretation of history. It seemed, he surmised, that Butterfield had sought to improve the tradition which he himself most shared. The book might better be entitled, said Smythe, *An Appeal from the Old Whigs to the New.*

Butterfield accepted the point at the time, and even today will talk of himself as neither a conservative nor a liberal, but as a life-long Whig—a New Whig. He prefers this term in order to qualify his relationship to the political and social tradition of liberalism:

> My Whiggism is different from liberalism in the continental or the American sense in that it is not utopian. Apart from

> hoping that human beings will be virtuous it does not operate
> by assuming that they are virtuous. My politics would operate
> by assuming that there is a great deal of egotism and cupidity
> in human beings—in *all* human beings, especially in those you
> say haven't got any cupidity.

The publication of *Whig Interpretation* in 1931 seems to mark the moment when Butterfield had settled the main features of his mature perspective. On the religious side, he had drawn his teenage Nonconformist beliefs and ethic into his Cambridge environment and decided for a religious direction which was neither liberal nor evangelical. As a historian who developed his talent and outlook among the debates and controversies of the 1920's, he had learned especially from Temperley, Trevelyan, Bury, Acton, and Ranke. He formed his own perspective as a New Whig with a love for the individual person, liberty, and tolerance compatible with his Nonconformity, and with an appreciation of history as a humble study possessing the marks of both science and literature. All allowances being made for a loose fit, it would seem that by about 1931 there are two terms available as guides to understanding his Christianity—English Nonconformist with qualities as a New Whig. It is important to determine the interconnections between his Christianity as understood in this way and his work as a historian.

Christian Historian

If it is the case that Butterfield's Methodist Nonconformity is the key to his Christianity, it is equally true that his Christianity is essential to understanding his work as a historian.

As a historian Butterfield has produced two large monographic histories in traditional fields of study. The first was *The Peace Tactics of Napoleon, 1806–08* (1929), a straightforward diplomatic history, 400 pages long, researched and written during the 1920's. It provided what is still regarded as the best explanation of the diplomatic relations between Napoleon and Czar Alexander surrounding the Franco-Russian Treaty of Tilsit of 1807: the Czar did not preplan a desertion of his Prussian ally, but, under the spell of Napoleon during their person-to-person encounter on a raft in the middle of the river, he allowed himself to be won over to Napoleon's side. After this book Butterfield wrote nothing more in diplomatic history, although he did publish on request a brief life of *Napoleon* (1939). He maintained an interest in the theory of international relations and wrote two books—*Christianity, Diplomacy, and War* (1953) and *International Conflict in the Twentieth Century* (1960)—and numerous articles on the subject. Many things combined to foster his interest in international relations, notably his training under Temperley, his attraction to Acton and Ranke, and, more personally, the shock of World War I and his need to understand why it had raged so furiously. He was particularly disturbed by the allies' desire to wage a "war for righteousness" to obliterate evil.

His second monograph was *George III, Lord North, and the People, 1779–80* (1949), another 400 page behemoth on a narrow time period. It was a study in political history, especially the interplay at a crucial conjuncture of extra-parliamentary opinion upon the king, the North ministry, and Parliament. "Our 'French Revolution' is in fact that of 1780—the revolution that we escaped," he wrote. The book became a standard in the field and, according to J. H. Plumb, "has shown how constitutional, political, and social history can be combined to achieve the reconstruction of a great crisis in our history."[4]

The subject connecting both monographs was the life of Charles James Fox. Even in the twenties he had conceived of writing a biography of Fox, but he backed away when he became aware that Trevelyan planned one. After *Whig Interpretation*, Trevelyan, who knew of Butterfield's attraction to Fox, magnanimously turned over to him the entire Fox Papers then in his possession, and Butterfield started in. He worked on Fox during the thirties and forties, spinning off articles here and there, as well as

a background book on Machiavelli (1940), and then *George III,
Lord North, and the People*. The last book drew him into controversy with the followers of Lewis Namier on the subject of
how to interpret George III, and eventually Butterfield dashed off
George III and the Historians (1957) as a reply. Occasionally he
published another monographic article or lecture on Fox—the
latest being "Sincerity and Insincerity in Charles James Fox"
(1972), but the biography remains to be completed.

Butterfield's greatest contribution as a historian has come not in
these more traditional fields, however, but in an area where he
has shown extraordinary originality—historiography and the history of historical study and writing. At least eight of his books,
plus his Gifford Lectures and this volume of his essays, fall within
this area. He has always been the kind of historian who wandered
off into questions of theory and method and framework and perspective in history. His first book, *The Historical Novel* (1924),
was an essay that defined how a historical novel was distinguished
from a book of historical scholarship. *Whig Interpretation* discussed the way historians viewed the past, and the method and
competence of historical study. *The Statecraft of Machiavelli*
(1940) was partly a study of Machiavelli's historical method and
framework of interpretation. *The Englishman and His History*
(1944) developed out of a request from German historians in the
1930's for him to explain the history of the Whig interpretation.
Christianity and History (1949), *History and Human Relations*
(1951), *Man on His Past* (1955), and *George III and the Historians* (1957) all belong to this field. In addition there are numerous articles and lectures on related themes, including many pieces
on his favorite historian, Lord Acton.

In some ways one of the tangent fields where he has exercised
considerable influence—the history of science—can be viewed as
an adjunct to his work in historiography. *The Origins of Modern
Science* (1949) is in one sense a lesson in how to apply to a specific
problem of history the theory of historical process and the ap-

proach to the historical framework which he expounded in *Whig Interpretation*. In articles he applied the same theory to the problems of toleration, religious liberty, and the history of Christianity.

Four of his books may be ranked as seminal works which have affected the way historians and a generation of history students have thought about history—*Whig Interpretation*, *Christianity and History*, *Origins of Modern Science*, and *Man on His Past*, and all four owe their originality to his ideas in historiography. Owen Chadwick, who succeeded Butterfield in 1968 as Regius Professor of Modern History at Cambridge, wrote about *Whig Interpretation*:

> [It] brought to an end an epoch of historical writing. It put historians into a state of self-analysis and scrupulosity. It was one of the two or three books from which stemmed the modern and fruitful consideration of the problems of historiography.[5]

Overall he has produced an unusual variety of types of writing in which he reveals himself as a historian. It ranges from his two very specialized and bulky monographs resting on his own original research to a number of encompassing general histories, like his masterful fifty page history of Christianity in *The Dictionary of the History of Ideas* (1973), and his as yet unfinished manuscript intended to be "The Cambridge Concise History of Modern Europe." He wrote as well some overview histories of more limited themes, like his life of Napoleon in 143 pages and *Origins of Modern Science*. He had provided over the years several statements on the historical method and the theoretical problems of epistemology in historical study, such as the first chapters of *Christianity and History* and *George III and the Historians*. In various writings he has examined the assumptions and approach of historians, especially in *George III and the Historians* on Namier, and in his reviews of books by major figures such as E. H. Carr, Christopher Dawson, Toynbee, G. P. Gooch, Pieter Geyl, Gor-

don Rupp, and many others. Then there are his productions in
the history of historical scholarship, notably *Man on His Past*,
and his two years of Gifford Lectures, as yet unpublished. Many
of his essays discussed what he called "historical-mindedness" and
the influence of ideas of history on the course of history. In other
works he presented his understanding of the structure of the
process of history and the laws in history, notably in *International Conflict in the Twentieth Century*. He has even written on
theory of international relations. He has given statements on theology of history, on meaning in history, and on Jesus and Israel,
especially in this volume of essays and in *Christianity and History*.
Perhaps the major type of writing he has neglected is autobiog-
raphy, which, he protests, he would find too difficult an under-
taking. By means of this variety, he provides us with explicit
entrée into his own outlook, approach, and practice as a historian
from top to bottom.

The striking thing about his *oeuvre* is that especially when
writing on the problems of historiography his work melts into
religious thinking. The years 1948 to 1956, which are the high
period of his thinking on Christianity and history, are also the important years in the expansion of his thought in historiography.
The union of the two areas is most complete in *Christianity and
History, History and Human Relations, Man on His Past*, and in
those essays in this volume coming from the same period. It is
plausible to understand his lecturing and writing on Christianity
and history as the point at which his professional vocation reaches
fulfillment and achieves greatest integration. For in this field of
labor he unites the three strongest vocational tendencies in his life
—the writer, the preacher, and the historian. Here he explicitly in-
tegrates what—at a self-conscious level at least—had perhaps
seemed to be two worlds—his inner world of personal religion and
his outer world of achievement as a historian. In the period from
1948 to 1956 he revealed unmistakably that his religious thought
depended on his experience and insight as a historian and that his

historical thought and writing rested upon the beliefs and ideas which he expounded as a religious thinker. To put the matter somewhat differently, it now became clear that his religious convictions underlay both his religious thought *and* his work as a historian. He is a Christian historian. Indeed, once he made explicit in his thought and writing the interconnections between his personal religion and his historical profession, it became clear that there was far more actual integration between his two worlds in the 1920's, 1930's, and 1940's than most people and perhaps even he himself realized. In other words, the historical work he published during and before 1949—including *Peace Tactics of Napoleon*, *Napoleon*, *George III, Lord North, and the People*, and *Origins of Modern Science*—in fact depended on his Christian perspective. The same observation would, of course, pertain to his work after 1949.

Before proceeding any further, it is important to mention that some disagreement exists among interpreters of the relation of Butterfield's religion to his work as a historian. There are people who read Butterfield's books and essays, even those from the 1948 to 1956 period, and claim that he keeps his religion out of his historical work. For example, Ved Mehta comments,

> What was remarkable was that, whatever Butterfield's religious views, they never colored his professional academic history, and, perhaps because he never hitched his lay history to the ecclesiastical wagon, he didn't forfeit his professional colleagues' respect or confidence.[6]

Any number of commentators believe Butterfield explicitly seeks to separate religion from history, by distinguishing between technical history as a limited mundane enquiry, and the Christian interpretation of history as a discourse on God and Providence. Patrick Gardiner notes the distinction as found in *Christianity and History*, but raises doubts about how successful he is in actually keeping them apart. "The two are different and, one is at first led

to suppose, unrelated," he writes.[7] Pieter Geyl is alarmed by what
he understands to be Butterfield's dictum in *Man on His Past* that
the two *should* remain on different levels and that historians
should learn to eliminate the God level in advance from the technical history level—thinking at different levels, in other words.
Declares Geyl,

> This is a view of history which I for one can never accept.
> History must claim the whole of life for its province. . . . I
> do not claim that history will solve the riddle of our existence.
> But the true historian . . . does not come to his material as a
> technician, but as a human being. He will eliminate nothing
> in advance.[8]

Karl Löwith willingly calls Butterfield "a wise historian and a
Christian," but insists that such a description is different from
saying he is a "Christian historian."[9] In reading Butterfield as they
do, these commentators may disclose more about themselves than
about him, and there is always the possibility that Butterfield's
statement of his theory may be inclined to mislead readers who,
in a secular and post-positivist age, are used to having religion
quartered in a house well away from history or politics.

The disagreements also may be due, partly, to how we set up
the problem. Butterfield certainly does desire to limit the domain
of the historian to those matters which the historical method is
competent to handle. In various writings between 1948 and 1956
he argues that historians are unable by academic research to discover the existence of God, the divinity of Christ, or the workings of Providence. One discovers such things, he says, by faith,
not by historical study. The method of history and the method
of faith are different, and cannot be otherwise. How does this
faith knowledge of God's Providence relate to the events of history that the historian studies? For Butterfield, in *Christianity and
History*, the believer may take the knowledge of history he ob-
tains by the historical method and survey it with a more penetrating look, thereby discerning the ways of Providence. Butterfield

tends to restrict the term "Christian *interpretation* of history" to any approach primarily focused on *God's* work and on the overview of history from the origins to the eschaton many find in the Bible. In this sense his religion and his professional history are separate.

There is another way to organize the problem, however. In his writings on Christianity and history, both in the period 1948 to 1956 and in his later work, including the essays in this volume, Butterfield identifies a whole network of ideas and beliefs which arise directly out of his religion and which serve as the fundamental instruments with which he approaches the study of human history. In this sense, he does not want to, nor does he actually, keep his personal religion separate from his professional history. To describe him as a Christian historian recognizes that his work as a historian is dependent upon and shaped by his Christian ideas and beliefs about human personality and freedom, redemption and human cupidity, spirituality and the mundane life, the dynamic of love and egotism in the historical process, the meaning and direction of the whole human drama, as well as the nature of truth, the validity of historical knowledge, and the limited, but worthy competence of any human action. Such beliefs have a very practical consequence for him in that they influence what he sees when he looks at human history. In this way what he learns in his religion about human beings he also tends to see in human history, just as what he observes in human history reinforces his faith. It happens that the constellation of ideas and beliefs underlying Butterfield's work are Nonconformist and New Whig, and they define his perspective as a Christian historian.

The Constellation of Ideas and Beliefs

We are faced with gathering Butterfield's ideas and beliefs from here and there throughout his career. Perhaps the best way to gain a comprehension that does justice to their character is to view his ideas historically, to see them as they develop and modify and interpenetrate. We may keep in mind that the years 1948 to 1956 represent a relatively stable time of fruition when he provides us with the next best thing to an overall statement of his beliefs at the height of his career. The discussion which follows will identify just a few of his leading ideas and assumptions, and indicate their relation to his religion and his historical work.

By far the single most important element in his thought is his idea of human personality in history. It is perhaps the most constant line of reference running throughout his corpus of writing. Butterfield would classify himself with historians like Carl Becker, Pieter Geyl, and G. P. Gooch whom he calls "essentially humanistic." Owen Chadwick even speaks of a "Cambridge school of historians," which Butterfield exemplifies, who, in contrast with some quantitative, Marxist, or behaviorist historians, emphasize that history is made by human beings. In his earliest work Butterfield was content simply to stress that history turns on what human beings do, and that people have wills and purposes that matter. The book to which he devoted most time in the twenties, *The Peace Tactics of Napoleon*, pivots on this point. In the preface he writes:

> The story has been told with special reference to the personalities engaged in the work of diplomacy, so that it might become apparent how much in these Napoleonic times the course of events could be deflected by the characters and the idiosyncracies of ambassadors and ministers who were far from home.

The book is full of superb portraits of individual personalities— Napoleon, Czar Alexander, Baron Hardenberg, Canning, even Fox. He laments the fact that so much diplomatic history is told as the operations of a system of diplomacy and institutions or of the logic of policy, and that historians can forget that human beings are at work. The central thesis of the study climaxes at this point:

> At Tilsit one can make no mistake. Here is the play of per-
> sonalities palpable and direct. It is not "Russia" that takes a
> course of action, like a piece of mechanical readjustment. . . .
> Everything is determined by personalities that act upon one
> another immediately.

For Butterfield the personalities are individuals possessing free
wills and vibrant souls and minds. He contrasts the "inner experience" with the "outer tangible world of actuality," the interior world of conversions of heart with the exterior world of
"mere incident and event."

In the context of Cambridge in the twenties, *Peace Tactics of
Napoleon* may be understood as an affirmation of the importance
of individual personalities over against the weighty emphasis
coming from Bury and the scientific historians on the power of
seemingly impersonal forces. Butterfield stuck with his Nonconformist beliefs in free will and the spiritual nature of the human personality. His diary covering his year at Princeton University in 1924–25 gives us a peek into his thinking and beliefs on
personality at the moment he was working on chapter two of the
book. He felt bombarded by the acquisitive materialism of American culture, and wished to emphasize the spiritual nature of life
and human beings:

> And yet, tho' economic needs and animal instincts are the
> sub-structure of life, the whole point of life is to raise an in-
> tellectual and spiritual synthesis on the top of these, so that
> these very things come out etherealised. Civilisation [is what]
> I would call this superstructure. It is determined first and
> foremost by material facts—by the fact that man is an animal
> who gets hungry and is moved by various instincts—but it
> gives these a spiritual significance and experience. (9 February
> 1925)

He added soon after, "The important thing is to turn the material
fact, the economic necessity, into a spiritual opportunity." A few
weeks later the local Methodist minister invited Butterfield to fill
the pulpit while the clergyman attended the annual Methodist

Conference in nearby Asbury Park. Butterfield wrote in his diary, "I felt too nervous, but I suppose I ought to do it if he can't get anyone to come from Asbury Park." The theme he chose for the spiritual edification of his congregation on Sunday evening, he recorded, was "On Personality," although he left no record of what he said. After preaching both morning and evening, he did note, however, "People seemed to be pleased with my services" (8 March 1925).

In *Whig Interpretation* (1931) Butterfield simply assumed and did not discuss his idea of personality. Instead he put his reconciling, dialectic method to work on incorporating what he could of the opposite point of view. The book stressed the reality of conditions and determining circumstances in history and commented on the supra-individual character of process in history. Individuals are entangled in "the web of circumstance," he said, and individual purposes are deflected by the process of conjunctures in history. For this reason, he affirmed, while we must always hold people responsible for what they do, we cannot issue the final moral judgment about them. We have already noticed his description of the process of history as a clash of wills, a conjuncture of acts and circumstances.

Butterfield encountered Marxism in the early 1930's, partly, he recalls, because so many undergraduates gave themselves to it. *Scrutiny*, in 1933, commissioned an article from him on the theme. His diary and notes during the thirties show him tangling with Marxism and communism, almost sounding as if he himself could convert to that position. What he did was to grant the essential truth of some Marxist theses which he thought Marxists overstated: that life is materially conditioned, that the process of history is dialectic, not linear, and that individuals are often overwhelmed by forces beyond their control. But, he affirmed all the more the importance of human personality—individual, spiritual, interior, free, and responsible. In his notes in the early thirties he urged,

> If we have an aim it is better that we should desire to rescue
> personality and open to men's minds a spiritual universe than
> to promote a political cause or a party programme.

His notes and published writings during the 1930's and early
1940's reveal him constantly ruminating on how to affirm the
primacy of human personality while yet granting the reality of
process and systems and determining forces. His biography, *Napoleon* (1939), depends on his view of human personality amid
the web of process and conditions. Then in the final chapter of
The Englishman and His History (1944) he declares poetically,
and with the ring of a confession of faith,

> Human beings, though fallen from the state of innocence,
> move as gods and bear the image of God; they are not part
> of the litter of the earth, to be left uncounted like the sands of
> the sea. Each is a precious jewel, each is a separate well of life,
> each we may say a separate poem; so that, without taking
> them in the mass, every single one of them has a value incom-
> mensurate with anything else in the created universe.

When he published *Christianity and History* (1949) he devoted a
whole chapter to "Human Nature in History" while also discussing process in history at length. It was in this book that he first
discussed one element that had always been an assumption of his
thought about the personality—the reality of human cupidity and
sin. For all the glory of the human personality, he was also impressed with the tendency in human beings toward idolatry, egotism, self-destruction, and self-righteousness. This "universal element of cupidity" in people he understood as a constant of history
in all ages.

His new monograph, *George III, Lord North, and the People*
(1949), like *Peace Tactics of Napoleon*, depended on his view of
human personality, and it reflected the subtle developments and
refinements of his thought on the theme: especially the subtleties
of the inner motives and purposes of a man, the universality of sin
on all sides of a struggle, the qualifying and entangling power of

process and conditions in history, and the importance of the historian's renunciation of the right of final moral judgment upon people. In the preface he stated how all these ideas of his affected his story:

> Everything is easy to the person who judges the Protestants in the ideal (or on their own evidence) and compares them with the Roman Catholics as they are in actuality (or as they appear on the evidence of their enemies); and many conjuring tricks are possible if we interpret one political party in terms of vested interests while construing the other in terms of its higher theoretical ends. The truth is that in politics the two are entangled—on both sides we are likely to find the play of vested interests, the sheer struggle for places, profits, and power, while on both sides there will be honest men, and some justification in terms of higher purpose. To see all these things together is perhaps a more difficult kind of history both to write and to read than many people realize; but it is this kind of history which—if we can get the combination into proper focus—will give us something like the stereoscopic picture, the landscape in proper relief.

Throughout the book Butterfield revealed unmistakably his sympathies for the Rockingham Whigs, especially Edmund Burke and Fox, and for the radicals like John Wilkes, all of whom, by their opposition, forced the situation open and moved the English political system along toward liberal parliamentary reform and a more democratic character. Yet even George III emerges from the story clothed in more impressive qualities than usual, and the radicals and Whigs appear to need the Tories in order to achieve a workable result.

Butterfield thought further about Marxism and made some comments on it in *Christianity and History*, a few more in the little essay produced about the same time, "The Christian and the Marxian Interpretation of History" (essay 12), and then more still in an essay entitled "Marxist History" in 1951. His renewed dialogue with Marxism, together with his study of the scientific revolution—*Origins of Modern Science* (1949)—inclined him to speak approvingly for the first time of "laws" in history. In his very first book, *The Historical Novel* (1924), he had specifically disparaged talk about "laws" as inappropriate for historians. But now he saw a valid use for the idea of law as an extension of the idea of conditions and process which he had expressed since *Whig Interpretation* (1931). In the essay "Marxist History" he concluded:

> In other words, we learn about men by learning how he is conditioned; and material conditions are no doubt reducible to laws in some degree, whether the Marxist has properly formulated the laws or not.

In "The Christian and Historical Study" (1951, essay 9) he takes the idea slightly farther in relating it to an overall idea of the created structures of the universe, what he elsewhere has called "the very constitution of the universe."

In 1952 Butterfield composed "God in History," the opening essay in this collection, and it remains the closest thing he has written to a synopsis of his views. It is a most original piece of work, for in it he offers a solution to the problem he struggled with for so long of how to reconcile synthetically and integrally his ideas of human personality, of conditions and laws in history, as well as the ancient belief he had made his own of Providence and God working in history. We can, he suggests, look at events in history and imagine them at three different levels of understanding and comprehend them with three different kinds of knowledge. All three understandings of history are true at the same time, and all three may be affirmed at once:

> We may say at the first level of analysis that men's actions make history—and men have free will—they are responsible for the kind of history that they make. But, then, secondly, at a different level, we find that history like nature itself, represents a realm of law—its events are in a certain sense reducible to laws. However unpredictable history may be before it has

> happened it is capable of rational explanation once it has hap-
> pened. . . . [T]here is a further factor that is operative in
> life and in the story of the centuries—one which in a sense
> includes these two other things—namely the Providence of
> God, in whom we live and move and have our being.

The first level we may study almost biographically; the second
we uncover through scientific, even statistical analysis of regularities and tendencies; and the third we approach by faith. The historian who brings his understanding of all three levels to his study
of any topic has the potential of enjoying a very rich and profound knowledge of human events. About any given event or
sequence, we might, therefore, make three simultaneously true
statements. For example, I came to the meeting tonight because I
decided to hear what the speaker had to say; or, almost all of us
who came to the meeting tonight were the type who can usually
be counted on to attend such a political discussion, so, I came because I am one of that type; or, I came to the meeting tonight
because God brought me, and I thank God for it.

It would be a mistake to think that Butterfield considers only
the last level of understanding as a Christian way of viewing history; it is here that commentators on Butterfield often go astray.
He regards the first two kinds of understanding as explicitly
Christian as well. Throughout his career, whether in his diary,
sermons, or historical writing, he has attributed the high yet balanced view of personality as both glorious and egotistical to the
Christian tradition in our civilization. More than once he has even
suggested that Nonconformity in particular has especially contributed to the modern form of the idea of personality in western
culture, and certainly to his own version of it. He makes a similar
claim about the influence of Christianity upon our culture's awareness of laws and regularities in the universe and in history. In his
own case he considers his understanding of laws in human history
as being not uncongenial with, and perhaps derivative from, his
Christian belief in reality as Creation made by the intelligent and
faithful Creator God.

Butterfield favors a presentation of history which, in a balanced and integrated way, makes use of all three levels of understanding. We see this in his debate with the Namierites. One of the main criticisms he brought against Sir Lewis Namier and his followers in *George III and the Historians* (1957) was that they dwelled too exclusively on the second level of understanding. Their analysis of the structure of politics was valid as far as it went, he suggested, but they left the events too much as if they were merely conditioned by forces and tendencies and types, and they lost hold of the range of choice that had been open to the people at the time. He added the criticism that when they did speak of personalities—the first level of understanding—they assumed too readily that politicians acted out of merely vested interests, and they neglected mention of the higher motives, particularly, as in the person of George III, "the framework of ideas and purposes which affect his actions and modify the course of things." Consequently they create a view of events as disconnected and characterized merely by chances and conjunctures. For Butterfield, in other words, history was best presented as an integration of free acts by personalities in the context of conditions which may in retrospect be described according to laws. And if the whole picture, without ascribing anything to Providence, could be colored by the sensitivity which comes from knowing that all rests in the hands of God, then how much richer the history can become.

Some crucial additional ideas and beliefs come into play in Butterfield's work when he writes about the nature of historical study and the history of historiography. As early as *The Historical Novel* (1924) he indicated his affinity with the Rankeian tradition of detailed documentary criticism. He was impressed, however, with the limitations of the documentary method—so many documents were lost, so much else is undocumented, and so much more is undocumentable. He remarked on "the impossibility of history," and, it seems fair to say, he even regarded the historical novel as not only different from historical study, but also superior to it. The novelist could fill in the gaps by his imagination and create the fuller picture, bringing the story nearer to the heart of human life. The limits of history, in this sense of the term, were due to defects or deficiencies in historical study.

By the time of *Whig Interpretation* he dropped the notion of the inferiority of historical study, and he was able to appreciate the limits of historical study in a different sense entirely. He now thought it was the virtue of history that it was competent to secure a specific type of knowledge within a particular range of life. The historian was the diviner of truth about the past, especially, at the first step, through the recovery of the unique detail, the concrete event, the external manifestations of personality, the contingent. The historian need not stop with the detail, however, but he may proceed to disclose the interconnections of the thing, to see it "entangled in the web of life." Like Ranke, Butterfield believed that every person, every individual thing, was valuable and near to God, and that all things were created, not in isolation, but in relationship with the others. The research to discover the truth as detail could yield the truth as general history. The historian can move beyond the externals, too, by employing the instrument of sympathetic imagination. This was an imagination, not to fill gaps and create fullness, as with the novelist, but to enter into the lives and personalities of by-gone people and to understand them as human beings, especially when they belonged to a different tradition or point of view from the historian. The historian was a reconciler.

In *Christianity and History*, "The Christian and Historical Study" (essay 9), and elsewhere, he shows that he regards the ideas about historical study which he laid out in *Whig Interpretation* to be essentially Christian ones, or at least compatible with Christianity—the concern for the truth, the value of the individual detail, the interconnectedness of life, the sympathetic imagination, the achievement of human understanding, the reconciling servant. He adds to these two others—elasticity of mind, and the exercise of charity and love. By elasticity, or flexibility, he means the practice of tolerance, the openness to new and different thoughts and people, the ability to rid ourselves of the unenduring and the false, the refusal to absolutize anything in this created world. By love and charity he means the Christ-like virtue of giving ourselves to others. Indeed, he says, love is almost the definition of elasticity itself, and for the Christian the law of love is the ultimate principle of life. The historian who shapes his method and approach with all these ideas and beliefs may enjoy the fullest possible vocation and still remain within the limits of his proper competence. He does not need to act as the moral judge, the prophet, or the pundit, and he need not feel cheated if, in his capacity as historian, he is unable to solve the riddle of life, or to explain the nature of things, or to provide the definitive commentary on the course of human events.

If for a moment we can hold in our minds what we have seen of Butterfield's idea of the three levels of understanding history, and join to that his idea of the method and competence of historical study, we will be able to approach more directly one of his most important passages on the relations between religion and historical study. It is the section in *Man on His Past* (pp. 139–41) that evoked Pieter Geyl's disapproval referred to earlier. Butterfield actually makes the same point in *Christianity and History* and elsewhere, but here he states his position concisely. It is often misconstrued to mean that he proposes an unqualified divorce of religion from historical study:

> The truth is that technical history is a limited and mundane realm of description and explanation, in which local and concrete things are achieved by a disciplined use of tangible evidence. I should not regard a thing as 'historically' established unless the proof were valid for the Catholic as well as the Protestant, for the Liberal as well as the Marxist. When Acton imputed to Ranke the design of laying out a demonstrable story, the facts of which were to be valid for men of all religions and parties, he may have been defining an impossible ideal, but he was defining what is really meant by techni-

cal history. . . . Each of these can add his judgements and
make his evaluations; and they can at least begin by having
some common ground for the great debate that still lies open
to them. Those who bring their religion to the interpretation
of the story are naturally giving a new dimension to events;
but they will not be less anxious than anybody else to know
what can be historically established.

Since producing this passage Butterfield has acknowledged that
very little—perhaps less than ten per cent—of what historians write
can be "historically established" according to his criterion. Moreover, although he has written less on this aspect of the subject, he
is aware that when historians move outward from the detail toward the fuller reconstruction of the episode, and even further to
the building up of the general framework, and when they move
inward by sympathetic imagination to the comprehension of motives and the inner character of personalities, the room increases
for the operation of religiously or ideologically based views of human nature and history. In other words, all parties might agree on
the date of George III's birth, but Marxists, Namierites, and liberals may, because of different ideas of human personality, disagree on what his motives were at the start of his reign, even on
whether his motives were relevant, and, more so, on where the
activity of George III fits into the larger political, social, and economic picture. It is also the case that Butterfield's own historical
writing is full of adjectival judgments of all sorts. The point is
that in this passage Butterfield is not talking about religiously
based views and beliefs about human nature and history filtering
into the story. On the contrary, we have seen that not only does
Butterfield's view of personality underlie his historical writing,
but also he himself thinks that it should.

What Butterfield asks for in this passage is that historians first
of all respect the limited competence of historical study, and hap-
pily restrain themselves from tendencies to self-aggrandize and to
absolutize historical study. He uses the terms "interpretation,"

"judgements," and "evaluations" to refer either to the third level of understanding of history as God's work, or to kinds of writing, like commentary, polemic, theology, or politics, which, while making use of history, are not properly historical writing as such. Butterfield hopes historians will perform the humble, yet invaluable task, of discovering the documentable detail and slowly moving with sympathetic imagination, elasticity of mind, and charity, to the enlargement of the story and the achievement of simple human understanding. In the face of skeptics and mere relativists, he wants to affirm the validity of historical knowledge, and in the context of the desperate conflicts of the last war and the post-war he recommends the priority of human understanding. One starting point, for him, is the small but nonetheless real common ground of what can be agreed upon irrespective of religion or ideology. As he wrote in *Christianity and History*, the historian may be the reconciler:

> Taking things retrospectively and recollecting in tranquility, the historian works over the past to cover the conflicts with understanding, and explains the unlikenesses between men and makes us sensible of their terrible predicaments; until at the finish—when all is as remote as the tale of Troy—we are able at least perhaps to be a little sorry for everybody.

There are many others of Butterfield's ideas and beliefs that might be discussed—such as his ideas about the social order, the structure of reality, the dynamic of history, the stages of civilization, the place of religion in history, the interpretation of the whole human drama, as well as his views of Jesus, the Resurrection, the Bible, Providence, and the hopes for Christianity in the twentieth century. But this discussion of his ideas of human personality and the process of history, together with a look at his assumptions about method, truth, understanding, and love in historical study, perhaps will suffice to show that throughout his career he developed a constellation of ideas and beliefs, specifically emergent from or compatible with his Nonconformist Christianity, which suffused his work as a historian.

The Essays

This volume of essays gathers together his most important articles and unpublished short writings on the theme of Christianity and history since his famous book in 1949. The essays are arranged in three parts. In Part I Butterfield discusses the role in history of God, process, the individual, the Christian religion, and moral judgment. In this section he articulates the main ideas about human nature and the historical process that he employs in his work as a historian. These are the ideas this Introduction has concentrated on. In Part II he first examines the sources of a Christian view of history—the Old and New Testaments, Eusebius, and Augustine. Then he suggests what insights are helpful for a Christian approach to historical study, and contrasts various interpretations and views of history. Here he states his ideas of method referred to in this Introduction. In Part III he assesses the place of Christianity in the twentieth century, and invites those who may share his beliefs to give themselves to a Christian life that is at once loving and tolerant of others, devoted to Christ, and open to the future.

In these essays Butterfield provides an amazing revelation of his personal religious beliefs and gives a suggestive commentary on the Bible, Jesus, the Resurrection, the short falls of Christians, and the opportunities open to Christians today. Especially if read together with his *Christianity and History*, what he offers in this volume is an invitation to "get behind the historian," as Lord Acton would say, and the means of doing so. Not many historians have provided us with so wide an access to understanding the ideas and beliefs and perspectives out of which they work.

Some things will be helpful to keep in mind while reading the essays. First, he always holds himself close to the happenings and the evidence of history. He has the mind and the knowledge of a historian. His special contribution to a discussion often dominated by theologians and biblical scholars is the outlook and experience of the historian. Second, he writes as a believer, a Christian believer who states his convictions as a confession of faith, and not as an intellectually refined doctrine. He does not write academic theology. Third, he is essentially an essayist who eschews theoretical systematization and comprehensive theory, and even analytic precision. His presentation is suggestive and highly impressionistic, and he often makes his point by means of the figure of speech.

Moreover, since the essays have been produced over nearly a thirty year period—1949 to 1977—there is a certain amount of overlap and minor repetition. He wrote for occasions, and he repeatedly consorts with a number of favorite ideas and themes. Even when he had moved on to a new development in his thought or interest, he would sweep back over a point he had made before. Words in his vocabulary which refer to his most fundamental beliefs—cupidity, wilful, mundane, spiritual, elasticity—serve him often. There is, after all, a remarkable continuity in his religion from his youth in Oxenhope until today.

If Butterfield has a message running through this volume it is the same one he has offered throughout his whole career: the supreme value in life is human personality. As a Christian he affirms that the spiritual character of personality is best fulfilled in communion with the Living Christ. The result can be "an insurgent Christianity" in the world. He closed *Christianity and History* with advice for meeting the future: "Hold to Christ, and for the rest be totally uncommitted." He ends this volume with the paradoxical formula for achieving insurgent Christianity: "give something of ourselves to contemplation and silence, and listening to the still small voice." And at such a moment in Herbert Butterfield, the historian and the religious thinker are one as he catches the word of God himself.

Notes

1. C. T. McIntire (ed.), *God, History, and Historians: Modern Christian Views of History* (New York: Oxford University Press, 1977).
2. G. M. Trevelyan, "Clio, a Muse" (1913), and "The Present Position of History: Inaugural Lecture" (1927), in *Clio, a Muse, and Other Essays* (London: Longmans, Green, 1930).
3. J. B. Bury, "The Science of History: Inaugural Lecture" (1903), and "Cleopatra's Nose" (1916), in *Selected Essays of J. B. Bury*, edited by Harold Temperley (Cambridge University Press, 1930).
4. J. H. Plumb, *England in the Eighteenth Century* (Penguin, 1950), pp. 216–17.
5. Owen Chadwick, *Freedom and the Historian: An Inaugural Lecture* (Cambridge University Press, 1969), p. 37.
6. Ved Mehta, *Fly and the Fly-Bottle: Encounters with British Intellectuals* (Boston: Little, Brown, 1963), pp. 251–52.
7. Patrick Gardiner, review of *Christianity and History*, in *Mind*, New Series, 60 (1951), 133–34.
8. Peter Geyl, *Encounters in History* (London: Collins Fontana, 1963, 1967), p. 256.
9. Karl Löwith, "History and Christianity," in *Reinhold Niebuhr: His Religious, Social, and Political Thought*, edited by Charles W. Kegley and Robert W. Bretall (New York: Macmillan, 1956, 1961), p. 290.

Bibliography
Writings and Lectures by Herbert Butterfield on Christianity and History

The list includes Butterfield's published writings germane to the theme of Christianity and history plus certain of his lectures. No book reviews are listed. Articles and lectures published in this volume are marked with an asterisk * and the essay number is identified in brackets [3]. In this way the reader may tell where and when each essay originated.

1931
The Whig Interpretation of History (London: Bell; and Penguin).

1944
The Englishman and His History (Cambridge University Press).

1947
"Reflections on the Predicament of Our Time," *Cambridge Journal*, 1.
"Limits of Historical Understanding," *Listener*, 37 (26 June).

1948
Lord Acton (London: Historical Association).
"The Protestant Church and the West," *Listener*, 39 (24 June).

1949
Christianity and History (London: Bell; New York: Scribner's, 1950).
*[10–13] "The Christian and Academic History," "The Christian and

the Biblical Interpretation of History," "The Christian and the
Marxian Interpretation of History," "The Christian and the
Ecclesiastical Interpretation of History," *The Christian News-
letter*, numbers 333 (16 March), 336 (27 April), and 314 (6
July).

1950
"The Cruxifixion in Human History," *British Weekly*, 127 (6 April).
"The Christian Idea of God," *Listener*, 44 (23 November).

1951
*[9] *History and Human Relations* (London: Collins; New York:
Macmillan, 1952); including "The Christian and Historical
Study."
Christianity in European History (Oxford University Press and
Collins).
"The Scientific versus the Moralistic Approach in International Af-
fairs," *International Affairs*, 27.
"Framework of the Future: By What Values?," *Listener*, 45 (22
March).
"The Contribution of Christianity to our Civilization," *Methodist Re-
corder* (3 May).
"Historian Looks at the World We Live In," *Religion in Education*,
18.

1952
Liberty in the Modern World (Toronto: Ryerson).
*[1] "God in History," Church of England Youth Council *Newsletter*
(July 1952); later published in Richard J. W. Bevan, (ed.),
Steps to Christian Understanding (Oxford University Press,
London, 1958), pp. 105–21.

1953
Christianity, Diplomacy, and War (London: Epworth).
"Prospect for Christianity," *Religion in Life*, 22.

1955
Man on His Past: The Study of the History of Historical Scholarship
(Cambridge University Press).
*[2] "The Role of the Individual in History," *History*, 40.
"The Christian and History," *Spectator*, 194 (29 April).

1956

*[14–16] Lectures on "Christianity in the Twentieth Century," Bristol
        University (January and February), including "The Challenge
        of the Faith," "The Obstruction to Belief," and "The Prospect
        for Christianity.

1957

Historical Development of the Principle of Toleration in British Life
        (London: Epworth), 17 pages.

1960

International Conflict in the Twentieth Century: A Christian View
        (London: Routledge and Kegan Paul; New York: Harper and
        Row).

1961

"Reflections on Religion and Modern Individualism," *Journal of the
        History of Ideas*, 22.

1964

Human Nature and the Dominion of Fear (Christian C.N.D. Pamphlet
        no. 3).

1965

Moral Judgments in History (London: Foundation Oration at Gold-
        smith's College, University of London).

*[7] The Gifford Lectures on "The History of Historical Writing,"
        Scotland, 1965 and 1966, including the original version of "The
        Establishment of a Christian Interpretation of World History."

1967
*[3] "Christianity and Politics," *Orbis*, 10.

1970
*[17] Address, St. Giles Cathedral, Edinburgh (19 April), later titled
        "Christians in the Coming Period of History."

1973
"Christianity in History," *The Dictionary of the History of Ideas*, II
        (1973), 373–412.
"Global Good and Evil," *New York Times* (3 and 4 January).

1974
*[4–6, 8] Lectures on "The Secular Historian and the Christian Reli-
    gion," Northwestern University (April); including "The Con-
    flict Between Right and Wrong in History," "The Originality
    of the Old Testament," "The Modern Historian and the New
    Testament," and the original version of "Does Belief in Chris-
    tianity Validly Affect the Modern Historian," which he revised
    and expanded in 1977.

I

The Divine and the Human
in History

I

God in History

Of all the factors which have operated to the disadvantage of religion and the undermining of the religious sense in recent centuries, the most damaging has been the notion of an absentee God who might be supposed to have created the universe in the first place, but who is then assumed to have left it to run as a piece of clockwork, so that he is outside our lives, outside history itself, unable to affect the course of things and hidden away from us by an impenetrable screen. This idea has had unfortunate effects upon religion and religious life even within the bosom of the Church itself. It has helped to discourage many people who were not really unfriendly to religion—helped to push them away on to the fringe of Christianity and into the position of friendly neutrals, sitting on the fence, but not quite convinced that there is anything very much that they can do about religion. Some of them believe in God but, since he is an absentee God, shut out from the events of this world, he might as well not be there—one can't really have a deep or a vivid kind of religion in relation to such a God. Indeed such a God does harm by giving people a comfortable general feeling of optimism about the universe, combined with a feeling that there is not very much point in troubling one-

3

self about religion. And if God cannot play a part in life, that is to say, in history, then neither can human beings have very much concern about him or very real relationships with him. Nothing is more important for the cause of religion at the present day than that we should recover the sense and consciousness of the Providence of God—a Providence that acts not merely by a species of remote control but as a living thing, operating in all the details of life—working at every moment, visible in every event. Without this you cannot have any serious religion, any real walking with God, any genuine prayer, any authentic fervour and faith.

It is clear that one of the reasons why people have lost their way in regard to this question is due to the effects of science, the effects of what I should call popular science rather than of the scientific mind or the scientific method as such. Partly this is the result of the fact that men have forgotten how the modern scientific method came into existence and the terms on which it came to be developed—have forgotten just what are the limits of what can be achieved by observing a blade of grass more and more microscopically or by looking at the stars through bigger and bigger telescopes. Partly people are over-awed by the things which we call "laws of nature," thinking of them as laws in the sense that Acts of Parliament are laws, when rather they are hypotheses—they represent our mode of understanding nature, our ways of formulating to ourselves the movements and processes that we observe to be taking place in the universe. And, if we make a mistake about this, our view of life and the world is apt to be very mechanical. In the old days they would teach musical composition by the use of mechanical rules, and the student had to do exercise after exercise until he had mastered the rules. The rules themselves were all taken from an analysis of the works of successful composers—it was possible to show that on the whole successful music did in fact conform to these particular rules. Everybody knew, however, that the great composers actually writing music never worte it to rule in this way—the rules were things which you discovered when you subjected good music to mechanical analysis afterwards. It would be a wild error to imagine that the composer created his music in the way in which students analysed it after it had been written. And similarly one must not imagine that God created the universe in the way in which we analyse it—more likely he resembled the composer who, we might say, was just out to create a beautiful thing. Above all one of the effects of this misunderstanding of the scientific method has been to give people a too mechanical and too abstract idea of God—one which fails to do justice to his fullness and richness. In particular we forget those significant words which St. Paul declared to the men of Athens when they, also, were worshipping an unknown God, too remote, too far removed from human life and history. St. Paul said: "He giveth to all life, and breath, and all things. . . . he is not far from every one of us; for in him we live and move and have our being." If we grasp those words properly and see all the world lying in the hollow of God's hand—see ourselves as living and moving only in him—then it becomes less difficult to imagine how intimately all Nature and History lie in the Providence of God.

Concerning the events that take place in nature and in history and in life there are three ways that we can have of looking at them—it might be said perhaps that we can imagine them at three different levels and with three different kinds of analysis. And because they are taken at different levels they can all be true at the same time, just as you could have three different shapes of the same piece of wood if you took three different cross-sections. If you go on a journey, and at the end of it I ask: Why are you here now? you may answer: "Because I wanted to come"; or you may say: "Because a railway-train carried me here"; or you may say: "Because it is the will of God"; and all these things may be true at the same time—true on different levels. So with history: we may say at the first level of analysis that men's actions make history—and men have free will—they are responsible for the kind of history that they make. But, then, secondly, at a different level, we find that history, like nature itself, represents a realm of law—its events are in a certain sense reducible to laws. However unpredictable history may be before it has happened it is capable of rational explanation once it has happened; so much so that it becomes difficult sometimes to imagine that it ever had been possible for anything else to have happened or for history to have taken any other course. Now these two things are difficult enough to reconcile in themselves—first of all the free will of human beings and secondly the reign of law in history. But they are reconcilable—and historians can discover large processes taking place in society for a hundred years to produce a French Revolution and an Industrial Revolution; and yet at the same time the historian will treat the French revolutionaries themselves or the nineteenth-century capitalists as subjects of free will, capable of making one decision rather than another, and even blamable for certain decisions that they actually made. We can even work out the laws and conditioning circumstances which have made the twentieth century an epoch of colossal world-wars; and those laws are so clear that some people were predicting their ultimate results nearly a hundred years ago. Some people in the nineteenth century, analysing the processes that were taking place in their time, predicted that the twentieth century would be a period of stupendous warfare and of still more prolonged war-strain. Yet, looking at the story from a different angle, we do not say that nobody is to blame for the outbreak of war in Europe in July 1914. The men who made disastrous decisions in July 1914 are still responsible and blamable for the decisions that they made.

But besides the freedom of the human will and besides the reign of law in history, there is a further factor that is operative in life and in the story of the centuries—one which in a sense includes these two other things—namely the Providence of God, in whom we live and move and have our being. And in part the Providence of God works through these two other things—it is Providence which puts us in a world where we run all the risks that follow from human free will and responsibility. It is Providence which puts us into a world that has its regularities and laws—a world therefore that we can do something with, provided we learn about the laws and the regularities of it. It would be wrong for us to picture God as interfering with the motion of the planets, stopping one of them arbitrarily, hurrying another of them along by sheer caprice—for we cannot imagine God as working by mere caprice. Indeed, centuries ago (before modern science had come into existence) men were looking for the laws and regularities in nature because they felt sure that God would not act by mere caprice. God is in all the motion of the planets—just as he is in all the motions of history. He is not interfering with the stars in their rotations—he is carrying them round all the time and in him they live and move and have their being. In his Providence he continues the original work of Creation and keeps the stars alight, maintains his world continually; we ought to feel that if he stopped breathing it would vanish into nothingness. It is like the case of the people you see in dreams—when you stop dreaming they no longer exist; and when God stops his work of creating and maintaining this universe we ourselves and all this fine pageantry of stars and planets simply cease to exist any longer. Those people are right who praise God every morning for the rising of the sun, and who see in this not merely a Providence which operated at the creation of the solar system millions of years ago, but evidence of his continuing care, his ever-present activity. It is not meaningless to praise God for the coming of the spring or for bringing us safely through to another day. It is because God is in everything, in every detail of life, that people so easily think that they can cancel him out. The world comes to do its thinking as though he did not exist.

Now that is the real affirmation that we as Christians have to make about life before the world at the present day—a world that is like something derelict and disinherited because it has lost touch with a really present God, with the real immediacy of God. It means that the Providence of God is at work in the downfall of Nazism, in the judgements that come on the British Empire for its own sins, in the present prosperity of the United States, and in our own individual daily experiences. That is what we see with the higher and more royal parts of our minds, when we make our highest judgements about life—our real valuations about events. And that is what we ought to say when we have our national joys, or our national victories or our national problems or our national dangers. We have to say: Providence has put us in this predicament—what can it mean? what moral good can we get out of it? what does God intend us to do when he puts this problem before us? what sins did we commit as a nation to merit this response from God and from history?

For let us make sure of one thing—in the long run there are only two alternative views about life or about history. Here is a fact which was realized thousands of years ago and it is still as true as ever. Either you trace everything back in the long run to sheer blind Chance, or you trace everything to God. Some of you might say that there is a third alternative—namely that everything just happens through the operation of the laws of nature. But that is not an explanation at all and the mind cannot rest there, for such a thesis does not tell us where the laws themselves can have come from. Either we must say that there is a mind behind the laws of nature—there is a God who ordered things in that particular way—or we must say that in the infinity of time all possible combinations of events are exhausted by the blind work of Chance, which produces amongst all the planets of the sky at any rate one where vegetation is possible and where animal life develops, and where in human beings matter itself acquires the quality of mind. There was one historian who outshone all others in regarding history as a science and historical events as subject to law. He was the Regius Professor of History in my University, a very famous scholar, Professor Bury, and in his Inaugural Lecture he stated in its most rigid form the scientific view of the course of history. But as he grew older he became greatly puzzled by the fact that he could explain why a Prime Minister happened to be walking down a street, and he could explain the scientific laws which loosened a tile on a roof so that it fell down at a particular moment; but he could not explain the conjuncture of the two—the fact that the Prime Minister should just be there to be killed by the falling tile—and yet it was just this *conjuncture* of the two things which was the most important feature of the story. What was more significant still—he found that all history was packed with these conjunctures—you can hardly consider anything in history without coming across them—so that this rigid believer in the firmness of scientific laws in history turned into the arch-prophet of the theory that Chance counted most of all. In his view the whole of the world's history was altered, for example, by the shape of Cleopatra's nose. Similarly he said that the Roman Empire fell in the West because of a handful of separate events which unfortunately happened to be taking place at the same time. Any single one of these events could be explained and reduced to law, but it was their conjuncture that mattered, and this Bury could only account for as the effect of Chance. Indeed when you go on analysing historical events further and further you find that the final problem of all—the really big thing that you have to solve—is this problem of the conjuncture. You can explain why each separate thing happened; but the important thing is the combination—the Prime Minister's path crossing the path of the tile at a given moment. If Hitler had been executed in 1925 or if Churchill had been killed in the Boer War we can be sure that all our history in this present year would have been different, though we should still have been able to work out scientifically the laws which help to explain how things came to happen in that way. Chance itself, or some equivalent of it, seems to have its part to play in historical explanation, therefore. And the historical process is much more subtle and flexible than most people seem to understand.

So, when we are considering historical events, there are three ways in which we can look at them. The first I would call the biographical way—we can see human beings taking their actions and decisions and operating with a certain amount of freedom so that they can be held responsible for the decisions they make; and in this sense men do make their own history and can blame themselves when their history goes wrong. The Christian would always have to be very emphatic about the free will of men and their moral responsibility—more emphatic I believe than anybody else; and he must come to the conclusion that all men are sinners—even all the statesmen of 1914 would have been wiser if they had had less egotism, less fear for their vested interests. The second way of looking at historical events is what I should really call the historical way rather than the biographical one—because it is the scientific examination of the deep forces and tendencies in history—the tendencies for example which had been making for war in Europe for fifty years before 1914, almost before the statesmen of 1914 were born—deep forces and tendencies which were working in fact for generations to help to make the twentieth century an era of colossal warfare. In this sense there is a part of men's history which the men themselves do not make—a history-making that goes on over their heads—helping to produce a French Revolution or an Industrial Revolution or a great war. And in this aspect of history we are much less inclined to blame the human beings concerned, and we see how much we have to be sorry for them. A Christian again must be most emphatic in his demand for this kind of history, this scientific kind of history, which examines the deep processes behind wars and revolutions and even tries to reduce them to law. And here is the great opportunity for Christian charity in history—here is why the Christian has to go over the past making no end of allowances for people—no end of ex-

planations—we might almost say that he cannot read history without being a little sorry for everybody. So you have free will in
history, and the statesmen of 1914 are blamable for unloosing the
horses of war. But also you have the operation of laws and processes in history; and the statesmen of 1914 are not as blamable as
they might have seemed at first sight, perhaps not more blamable
than you yourself might have been if you had been in the same
historical predicament—perhaps not more blamable than you yourself have often been at moments when the disaster was only reduced because you did not happen to be a statesman responsible
for the welfare of millions of people. Thirdly, however, you have
to think of another aspect. Either you must say that Chance is one
of the greatest factors in history and that the whole of the story
is in the last resort the product of blind Chance, or you must say
that the whole of it is in the hands of Providence—in him we live
and move and have our being—even the free will of men and even
the operation of law in history, even these are within Providence
itself and under it. But if you say that it is Providence, you must
not imagine that Providence can act merely in a chancy and capri-
cious way—Providence is acting in all that part of history which
is subject to law as well as in all that part of history which men
otherwise tend to attribute to Chance. And if you hold this view,
then there is a further way of looking at the war of 1914—you
must regard that war as itself a judgement of God on certain evils
of our civilization which could not be rooted out in any other
way. And if you look at the question in this light you can even
discover what those evils actually were. Indeed we know what the
moral diseases of the pre-1914 world were, which led to the outbreak of a European war.

Of course it is possible to read history and study the course of
centuries without seeing God in the story at all; just as it is possible for men to live their lives in the present day without seeing
that God has any part to play. I could not go to people and say
that if they studied nearly two thousand years of European his-

tory this would be bound to make them Christian; I could not say that such a stretch of history would prove to any impartial person that Providence underlies the whole human drama. You can learn about the ups-and-downs of one state and another in one century and another, you can learn about the rise of vast empires and the growth of big organizations and the evolution of democracy or the development of modern science—and all this will not show you God in history if you have not found God in your daily life. When we seek to know how God is revealed in history we do not make a chart or a diagram of all the centuries and try to show to what future great world-empires are tending or to what end great human organizations are moving. Russia, the United States, England—these are only names on a map, and if we know anything we know that some time in the far future men will be asking what was this thing called England, just as we ask about Assyria and Tyre and Sidon—some day the archaeologists will be rummaging amongst the ruins of London just as we excavate for Nineveh and hunt for ancient Troy. If we wish to know how God works in history we shall not find it by looking at the charts of all the centuries—we have to begin by seeing how God works in our individual lives and then we expand this on to the scale of the nation, we project it on to the scale of mankind. Only those who have brought God home to themselves in this way will be able to see him at work in history, and without this we might be tempted to see history as a tale told by an idiot, a product of blind Chance. If a great misfortune comes on us we may just feel how unlucky we are when compared with all our other friends who had previously seemed to be in a condition similar to ours. We need not adopt this attitude, however; there are some people who bring their sins home to themselves and say that this is a chastisement from God; or they say that God is testing them, trying them in the fire, fitting them for some more important work that he has for them to do. Those who adopt this view in their individual lives will easily see that it enlarges and projects itself on to the scale of all history; it affects our interpretation of national misfortunes as well as private ones. And when we reach this point in the argument we realize that we are adopting the biblical interpretation of history.

The way that God reveals himself in history is in fact the great theme of the Bible itself. And if you want to ask: "How does God reveal himself in ordinary secular history?" then it is exactly this which is the particular theme of the Old Testament. The Old Testament is the history of a people whose fate and vicissitudes were uncommonly like those of most other states—even modern ones. If the history was peculiar, it was perhaps in being worse and more violent than that of other states; for the ancient Hebrews lived in a tiny country with vast empires rising on either side of them and they retained their political independence only for a moment, only for a tiny fraction of their history. Afterwards, down till the twentieth century their land remained under the heel of vast empires, the Assyrian, the Babylonian, the Persian, the Roman, the Arabian, and the Turkish Empire in turn. Where they differed from other nations was in the way in which they interpreted their history—in fact it was their way of interpreting their history that was their chief contribution to the development of civilization. Because of that, they are remembered today and hold a high place in the world's story, even though they were no bigger than Wales and retained their political independence for so short a period.

They saw God as being essentially the God of History, and the result was that first and foremost they regarded history as based on the Promise. And although they took this Promise in a purely nationalistic way, all Christians must regard history as based finally on the Promise—it is never permitted to a Christian to despair of Providence. But the Children of Israel sinned, and theirs is the only national history I ever remember reading which proclaimed the sinfulness of the nation—proclaimed its own nation even to be worse than the other pagan nations round about them. And because of this, history at the second stage of the argument appeared to this people in its aspect as Judgement. When colossal national disaster came upon them they saw the tragedy as the effect of a Judgement from Heaven. At the next stage, however, they saw that God's Judgement does not cancel his Promise—if God judged the nation it was only in order to save it—for God is Love and it is always dangerous to think of the power of God without also thinking of his love. Even when their distresses were at their greatest and God seemed to be chastising them most severely, they came to what I think is the ultimate picture of God in History—God looking upon this world of cupidity and cross-purposes, of violence and of conflict—and pulling upon it like a magnet—drawing men with his loving-kindness. Judgement might fall heavily upon them but they were undefeatable in one respect. They saw that the Judgement did not cancel the Promise.

The Children of Israel had actually come in sight of the Promised Land—their spies had actually entered it and brought samples of its rich fruits—when they met an unexpected enemy whom God told them they must fight before they could actually enter the Promised Land—and they rebelled against him for putting them to this further trial. God brought us out of the land of Egypt because he hated us, they said—he brought us here only to entrap us. Let us go back to the land of Egypt, back to the House of Bondage. Let us make a captain and let us return to Egypt, they said. And God was so angry that he said they should not see the Land of Promise—their carcasses should fall in the wilderness. But in spite of the Judgement he kept his Promise to the Children of Israel—for though these people themselves were not to enter the land of Canaan, he decreed that their children should come into it later. He did not take hope away from the world.

And much later than that, when in the days of the prophets Judgement came upon the Kingdoms that the Children of Israel had established, and Jerusalem itself was razed to the ground—still once again the Judgement did not cancel the Promise—God said that he would make a new covenant with his people. He said that even his sending them into exile was meant for their further good—and we know that these particular experiences deepened the religion of the Children of Israel in many remarkable ways, so that this period of defeat and anguish was one of their great creative moments. Just at this time they gave a new development to the history of religion and religious thought. It proved to be an immortal moment for them.

But even now they sinned through excessive nationalism and worldly-mindedness, and when God made a new Covenant with them, they took it to mean a promise of new worldly success, victory in war, glory for their kingdom, dominion on earth. After this date the ancient Hebrews committed some terrible and wilful mistakes because they believed that God was to be on their side in battle and was to bring their nation to the top of the world—they even believed that the promised Messiah was to be a warrior-leader, a military saviour. Again their punishment was terrible and tragedy after tragedy came upon their endeavours; but the Judgement did not cancel the Promise. The trouble was that God's Promise to them was a higher thing than they knew, a better thing than they had imagined. They had been construing God's Promise in too worldly a way. And, though they were wrong, the Promise was not cancelled.

The greatest of the Old Testament Jews came to realize that God's Promise was not one of luxury and worldly success—it was a Promise that the nation should have a mission—a mission that should give meaning to its very disasters—and it was through that mission that it was to have a great role in history, an immortal name amongst men. Its mission was to teach the nations of the world about God—to spread to the rest of mankind the special revelation it had had—the knowledge of God as revealed in history. Through the very sufferings of Israel the world was to be carried to a higher religious life. It was the final mark of their mission that Christ was to come into the world as a member of this stricken, oppressed subject-nation—a Messiah who was to bring them greater glory and fame, a greater place in history than any warrior-leader ever could have done. It was a sign of the blindness of even this nation at the crucial moment that it rejected the Messiah when the Messiah actually came. That is the history of a nation whose stories of violence and conflict, treachery and war, of worldliness and cupidity, could be told just like the story of any other nation. The one difference was that the ancient Jews interpreted their history differently and saw the hand of God in it—they had the same experiences as other nations but they turned those experiences into spiritual experiences and because of that their history did become different—because of that they achieved creative things. For the greatest triumph of spirit over matter is when people can turn even their defeats and distresses into a creative moment like that.

We must imagine Providence as doing the best that the wilfulness of men allows it to do. For all of us History is the Promise and we need never despair—but it is a Promise punctuated by acts of Judgement. Even the great disasters of history, like the Jewish Exile, or the downfall of the Roman Empire, or the Norman Conquest of England can turn out to appear in history as a colossal benefit to mankind, and Providence can draw even good out of evil. It can even use our past sins to serve its future purposes. The Judgement of God may come upon an old world only to make way for a new one. Perhaps it is the only way in which on occasion the world in general can be induced to rise higher.

2

The Role of the Individual
in History

I

It would appear that, whatever other things may be superadded, the primary interest in history is the ordinary human interest—the desire to know about our predecessors, to keep certain episodes alive in the memory, and to recapture the past as both a picture and a story. If we were to start with an historical blank, wiping out all the existing literature of the subject, exploding all the reasons for teaching it—if even we were to decide that historical knowledge itself were fallacious and dangerous—I doubt whether we could stop children from listening to the tales of a grandfather, or deter people from inquiring about Stalin, or prevent the retelling of the story of Operation Cicero. It is possible to wonder whether the development of scientific techniques may not end by transforming academic history itself into something like a species of algebra. We can be sure, however, that the thing we ordinarily call history—the history which is one of the humanities and which merely recovers and narrates—would persist if only in the hands of the popularizers, rising still, at times, into the realm of enduring literature.

Having this primary kind of history in mind I should like to take as the starting-point of an argument certain theses which I have already stated in print; but as the arteries harden and the mind freezes into rigidities I do not quite know how to find fresh words for these. It will perhaps be permissible for me to repeat them:

> The genesis of historical events lies in human beings. The real birth of ideas takes place in human brains. The reason why things happen is that human beings have vitality. From the historian's point of view it is this that makes the world go round. If we take all the individuals of France at a given moment in 1789, they represent what in one respect can be regarded as so many separate wells of life, so many sources of decision and action. . . . If we . . . think of the French Revolution as a 'thing' . . . above all, if we start imagining that the French Revolution stood up and did something as though it were a self-acting agent (when we really mean that a certain man or group of men came to some decision or other) . . . then we are moving into [a] world of optical illusions. . . . Economic factors, financial situations, wars, political crises, do not cause anything, do not do anything, do not exist except as abstract terms and convenient pieces of short-hand. . . . It is men who make history.

Regarded from this point of view history is an intricate network formed by all the things that happen to individuals and all the things that individuals do. In other words it is "the essence of innumerable biographies"; and even the "history of thought" may lead to deception unless we regard it as rather the history of people thinking. In economic history the human beings may be too numerous to be mentioned by name; and in the history of ideas we must often doubt whether we have discovered the original inventor of a new thing. But as Carlyle said, the man who first took an army over the Alps is not more momentous to us than "the nameless boor who first hammered out an iron spade." In one way or another everything in history is ultimately referable to individual people.

Because we can never fully know the internal life of another human being, there is a sense in which personalities are irreducible entities for the student of either the past or the present. Because we can never know the remoter workings of mind and motive in other people, there is a sense in which we can never get behind the fact that Napoleon at a given moment made a certain decision. We may try to collect all the factors that help to explain the decision or produced the conditions for it, but still there is something left over; for the historian has no data which would authorize him to prejudge the question of human free will and responsibility in contravention of those principles which he must follow in his commerce with his own contemporary world. I can say to my next-door neighbour: "You might have maimed that carol-singer but you ought not to have killed him." But if I can regard my neighbour as having the power to choose his action I must also assume that Napoleon over and over again might have altered the course of history by making a choice different from the one which he actually made.

This means that there is something in history to which we can do justice only by reproducing the course of events as a story, the kind of story in which we do not know what is going to happen next. And most of us are aware in fact that the understanding of the past may be obstructed if the student has been unable to unload from his mind the foreknowledge that he possesses, his awareness of how the story is going to end. When the narrative is allowed to present itself in hard lines, giving an impression of rigid inevitability, such an effect is calculated to make us sceptical of the possibility of altering the world by any action of ours. It is important, therefore, to remember that the effect is a trick of the historian's mirror; for there is no irrevocability in human action except that which is the equivalent of the statement that the action has already been performed.

Dorothy Walsh, writing in the *Journal of Philosophy* a few decades ago, pointed out that since men make choices there is something more than mere process in history—something over and above the forces, factors and tendencies that we so often discuss. History, she argues, therefore, is the revelation of what man has added to creation itself. One thing is clear: we are not entitled to imagine that the past ever quite determines or explains the future; for the wills and choices of human beings here in the present are always interposed between the two. Nor would it be relevant to argue that "the men who decide the turn of events at a given moment are themselves only the product of their age." Of two products of the same manse the one became a nonconformist minister because his father was a nonconformist minister; the other became an enemy of religion because his father was a nonconformist minister. It has been suggested that the different reactions of two such brothers may show the effect of a cat jumping into the cradle of the one of them before he was six months old; but, however rigid a determinist I may be, I can hardly escape an extremely flexible view of the transition from one generation to another. We can collect everything that preceded Shakespeare and imagine that we hold all the possible ingredients in our hands, but until we have that original fountain and source of new things— the mind of Shakespeare himself—we cannot pretend that we possess any equivalent for his plays. I ask to be allowed to repeat what I have said before: namely that the influences and ingredients which an age or an environment supply are

> churned over afresh inside any human personality, each man assimilating them, combining them and reacting to them in his peculiar way. The result is that nobody is to be explained as the *mere* product of his age; but every personality is a fountain of action . . . capable of producing new things.

Since individuals are presumed to have a region within which they make choices, their decision on any given occasion can never be merely inferred any more than it could have been predicted;

it can only be discovered empirically. It is in keeping with this fact, I think, that the genuine historical mind always hankers after concreteness and particularity. Minute researches, concentrated on the action of individuals day by day, are the solid rock on which historical scholarship is built. Bergson once wrote:

> I am inclined to believe that the rôle of fate has been exaggerated. Not enough attention has been paid to the very great importance of small, very small, accidental circumstances.

I should like to underline Bergson's remark, for I do not see how any reader of general history who enters at strategic points into the details of the story can avoid being impressed by the displacements which can be produced by some microscopic circumstance that initially might have seemed irrelevant. And those who stress the rôle of the individual in history are *ipso facto* committed to recognizing the importance of contingency. Listening to a passage of musical recapitulation where an expected B suddenly becomes altered into a B flat and a whole mighty orchestra thereupon swings round into a new orbit, one cannot help recalling on how small a pivot the grandest kind of history sometimes turns. It would be wrong to picture the activity of personalities as a mere ornament or a descant or violin obligato—as a mere playing on the fringes of great historical processes; for the processes themselves take place only through people, and the work of a Galileo or a Napoleon or a Lenin cuts into the whole structure of history. Furthermore, if Hitler had been executed in 1925 it is not just the case that "the other thing" would have happened, or that the alternative future which we dream of would then have been actualized. It is millions of men who would have chosen different things and would indeed have had vastly different ranges of things to choose from; and we cannot unravel the changes or count the alternative futures that might have been, any more than we can predict the multiple possibilities that are open to us at the present moment.

All the intricate tracery of the naked twigs on a wintry tree, therefore, and all the subtlety with which they grade into the larger branches, are insufficient to suggest the delicacy of the interplay in the field of history, or the obliqueness of the strategies involved in large historical change. As we have no notation with which to express the flexibility of historical development and the possibilities of surprise that lie in the passage from one generation to another, it will perhaps be permissible to offer a formula which might serve as a parable, and which typifies the truth if only by presenting a limiting case. Having been a spasmodic reader of history for a long period and having noted what leverages and transmissions of force occur when anything gets into the machinery of the historical process, I find myself more and more confirmed in the view that what we understand by a change in the course of world-history could be produced through the conscious purpose of twenty men, none of them possessing artificial advantages at the beginning of the story—twenty men united by a sense of mission.

Even when we are dealing with humanity in large masses or discussing some great event in terms of "general causes," this impersonal technique should never blind us to the fact that live human beings are in question. And the behavior of each of these does not cease to matter however multitudinous the crowd they may be in. Carlyle, discussing the responsibility for the condition of France on the eve of the Revolution, tells us that some men blamed Turgot, others said it was Necker, others said it was the queen; they argued "it was he, it was she, it was that." Carlyle caught hold of one side of the truth when he said that every man who had done less than his duty had contributed to the evil—had brought his thread to the production of that piece of historical tapestry.

> Friends! it was every scoundrel that had lived, and quack-
> like pretended to be doing, and been only eating and *mis-*
> doing, in all provinces of life, as Shoeblack or as Sovereign
> Lord, each in his degree.

Even today, when the affairs of a country culminate in a gigantic offence, we have only to stare at the process of things for a little while to see that, behind the shocking crime at the centre, the blame goes back to vast numbers of individuals, each guilty of small derelictions of duty or petty compliances with vested inter-ests—each gravely responsible, though astoundingly unaware of the importance of what he was doing. And one of the difficulties of modern democracy is not that individuals matter less than before but that they tend to think that they matter less. The responsibility is so dispersed that each man can close his eyes to the significance of his own part in the events that take place. One can fall into what one imagines to be a microscopic delinquency, in the secret hope that everybody else will not build it up into a giant evil by doing the same. Or one may say to oneself that since everybody will be taking this particular liberty one cannot hope to do any good by swimming against the tide. Carlyle's thesis carries the consequence that wars may be caused, or empires fall, or civilizations decline, not necessarily through some colossal criminality in the first place, but from multitudinous cases of petty betrayal or individual neglect. It is conceivable that it would have required no great change in human nature generally, but only a little less wilfulness in great masses of people at one time and another, to hold in check some of the more monstrous evils of the twentieth century.

The levers and pulleys in the historical process work so trickily that all of us must be able to recall occasions when our doing a trifle more or a trifle less than our duty has had a magnified effect which we should never have calculated. Not only do little things become magnified, but big things, like our victory in the Second

World War, sometimes appear to be achieved by so small a margin that we hold our breath at the memory of the hairbreadth escape. There is generally an opposition to liberal measures in any country, and even in England that opposition has often been strong enough to secure delay. It is not clear that a colossal margin is needed to decide on a pivotal occasion whether a given country shall be switched into a liberal or a reactionary course. In the first volume of his large *History of Science* George Sarton warns us not to imagine that the Greeks were a race of scientists— he points out that, in the main, the masses were really religious. The great prophets of Israel appear at the same time to have been small in number, preaching often, on uncongenial soil. If the ancient Hebrews are remembered as having advanced beyond their neighbours in religion, while the ancient Greeks showed their predominance in science, there would seem to be no reason why this should be imputed to a national mentality or a geographical determinism. A handful of peculiar men in each of the countries concerned might account for the divergence in the historical development and historical rôle of the two peoples.

If it is the individual who matters most in the sense that he is the maker of history, the next important force and the strongest organizational unit in the world's story would appear to be the thing which we call a "cell"; for it is a remorseless self-multiplier; it is exceptionally difficult to destroy; it can preserve its intensity of local life while vast organizations quickly wither when they are weakened at the centre; it can defy the power of governments; and it is the appropriate lever for prising open any *status quo*. Whether we take early Christianity or sixteenth-century Calvinism or the French revolutionary period or modern communism, this seems the appointed way by which a mere handful of people may open a new chapter in the history of civilization. And the men who form cells are pursuing a higher strategy than those who seek immediately to capture governments; for those who make a direct bid to capture a government must bow before existing gods and existing tendencies in order to open a path to power; while those who form cells have no need to dilute their purposes or to purchase favour from the supporters of the *status quo*.

In all this we have been envisaging the past which the narrative historians try to resurrect—just the spectacle of the life of man on the earth, the cinematograph film which ends in our present. And, like the narrative historians we have been considering past actions as though we were back in its own present, at the moment when it could be seen in its fluid state, and before it had been frozen into inevocability by virtue of being an accomplished fact. Only in this way only by talking about history as about life on the earth in general—can we really discuss whether individuals make history or merely go where history takes them. We have to start the argument from the assumptions on which we conduct our lives at the present day, and we cannot discuss the matter without constant cross-reference to the life we actually experience in the present. When past and present are taken together in an over-all view in this way it seems to me that the individual matters in the whole story much more than he tends to imagine, and matters even when he feels himself to be submerged in vast crowds. He matters all the more in that history represents a field in which big decisions can be carried by a narrow margin. The individual's actions produce perhaps a disproportionate displacement when he goes against the crowd; and indeed a small number of such exceptional cases may prove so pivotal as to suggest that God might have saved Sodom for the sake of ten righteous men without doing any violence to the historical process. I do not know how to answer those people who argue that if Napoleon had never lived the capitalist system would have developed exactly as it has done. I think such a system capable of any number of developments, any one of which will appear logical when one has come to the end of the story. I should be very suspicious of the view that physics and mathematics have so to speak a pre-determined course of development, and, for example, I disagree with those who assume that the modern necessity for concentration on war does not affect the essential evolution or direction of science. But even if we grant that some things take a rigid path, the over-all course of historical development certainly does not. It is capable of being switched into one of a myriad of alternative courses in accordance with what men decide at every moment of their lives. World-history in the aspects of it which are momentous for men's lives takes a great new turn from the mere fact that Hitler has lived and it would have taken another vastly different turn if he had been victorious in the Second World War. And the Providence which does not overlook the falling sparrow leaves something of the larger destiny of nations and civilizations for individual human beings to decide.

II

Yet the margin that we may leave for choices made by individuals in conditions under which some other choice is presumed to have been possible for them is like a small segment cut out of the great circle of necessity. If any such margin is granted at all it is sufficient to alter the character of all history, just as one single freak of chance or a single coincidence can transform a whole story. The principal debate in this field would be raised by those people who think it folly to imagine that any margin of genuine freedom can exist at all. None would challenge the ascription of a considerable area of human life to the sphere of law and necessity.

And history adds emphasis to this aspect of the matter too. It suggests that in one dimension at least the range of choice and effective decision open to human beings, and especially perhaps to statesmen, is more constricted than is often imagined, while in their very acts of choice men are more highly conditioned, more the slaves of their own past, than they are aware. Even in the development of his personality each is more committed to the given ingredients (and more subject to the state of the weather in the use he makes of these) than is apparent at first view. A man may be responsible for the kind of lover he is, but all outside observers are aware that only the icy determinism of the inscrutable stars decides on what woman he shall fix his love-making. I have often heard it said that a man who seemed constituted to be a rake has sometimes shown the ability to be a rare kind of saint, and I believe this to be the case; though there seems to be some doubt as to whether such a man can ever find a lodging at any half-way house. Napoleon, having made the kind of master-choice which so often governs further decisions (the series continuing into an indefinite future) tried on many occasions to pretend that he presided imperially over events. Yet there leaks into some of his utterances the consciousness of being the prisoner of a relentless destiny. For all these reasons it is relevant to examine another approach to history which is rather an analysis of the kind of necessity under which the individual lives.

If we roll into one the whole of our experience direct and indirect, seeing the past as continuous with the present, and uniting that appreciation of human nature which comes from our self-knowledge with that which comes from observation—this all-embracing view of the human drama clearly presides over our general outlook. There is a different view which we can take of the past, however, and this, though it should never have the same presiding place—should never be decisive of our over-all outlook, because it does not arise from an over-all view—is a thing which it would be imprudent ever to reject. It ought not to undermine that principle of individual responsibility which we have accepted at the higher level of analysis. But it produces great displacements in our views about the points where the responsibility lies.

Towards the end of the seventeenth century the world became more conscious and definite than before in its awareness of the scientific results which could be achieved by taking observations of external events and by making correlations between these. A more specialized form of thinking proved possible if one noted the

behaviour of pellets on an inclined plane without troubling about
the internal life of the pellet or the urges and aspirations that
might be within it; or if one meditated on the mere quality of fall
in the falling apple, without reference to what the essential nature
of an apple might be. Since so much of our thinking always
depended on the correlations made between observed external
effects—and even the practice of agriculture could never have developed without it—the procedure can hardly be regarded as having been a new one; and it was only by virtue of a certain intent-
ness of specialization that the seventeenth century gave the story
a significant turn. New issues were bound to be raised, however,
if the same procedures were transferred to history and human affairs—the study of the externals of observed events and the making of correlations between them, irrespective of the insides of
the people who might be concerned in the events, and ignoring
the element of personality.

Yet the eighteenth century, taking conscious advantage of the
prestige of Newton, gave a considerable development to this
method even in the field of human studies. The basic procedure
was simple, and of course it was only the more specialized and
conscious use of it that was new in the Age of Reason. Machiavelli, under the influence of classical Greece, had illustrated the
method by going through all the conspiracies of history and trying to discover whether—irrespective of personalities and inci-
dentals—one could not learn the pitfalls to be avoided by men who
wanted to assassinate their king. His project comes to a climax in
the Marxian attempt to study all revolutions for the purpose of
learning the science of revolutions—how to time them, and how
to know what are the strategic moves to make, for example. David
Hume had some faith in this attitude to historical data—consciously attempting to follow Machiavelli on the one hand and
Sir Isaac Newton on the other. He produced a thesis which was
calculated and perhaps actually intended to open a wider door for
this kind of human science. He claimed to demonstrate that the

existence of virtue in public life was not dependent on the private virtue of the individuals concerned but corresponded to the existence of well-regulated institutions. He sought to show from ancient history that periods during which men were virtuous in their private life could coincide with the absence of public spirit, and *vice versa*. He thought to clinch his case by comparing the corruption which existed in the ill-regulated republic of Genoa with the reputation for distinguished integrity which the self-same individuals enjoyed in their management of the Bank of St. George—a well-regulated institution. He had certainly seized on a genuine discrepancy here, for many of us have probably known men who were even most self-sacrificing in all the private walks of life, but seemed to have a different political self and were capable of great cruelty in public affairs, especially against people whom they never actually met in the flesh. This whole method deals with men in their aspect as moulded by circumstance, as creatures manufactured by history, rather than as the makers and choosers of their course. Some high ecclesiastical authorities in the twentieth century have gone further than Hume and have asserted that Soviet Russia is engaged in the task of producing a revision of human nature itself to suit its own purposes. I suspect that if such an achievement were in any radical sense possible, it must have happened in the world long before the twentieth century, so that all of us must already be victims of it. But if we imagine it to be achieved there would be some delicate calculations to make concerning the moral responsibility of those who lived under the new conditions; for the whole hypothesis is hardly consistent with the view that men are entirely responsible for their own dispositions or their own nature.

The same method has sometimes led to the study of the past not as the story of individuals but as the examination of the processes taking place within a society, the analysis of the rise and fall of civilizations, the comparative study of religions. Indeed all the modern sciences of society would seem to spring out of this

alternative use of historical data, and by virtue of it we come to
see in history profound processes or sweeping tides which move
forward as though more powerful than the wills of particular
men. It seems possible, therefore, to assemble a kind of history in
which individuals as such do not seem to possess significance.
Methods of abstraction are involved in this treatment of the past
and one no longer sets before one's eyes the whole complex
variety of the world—one no longer considers individual people—
but one picks out special kinds of data for correlation. One therefore cannot base one's more general reflections about life on this
kind of history, and if one deduces from it that the individual
really does not matter is history one is clearly arguing in a circle.
Here is so to speak a pocket of human thinking—a pocket into
which one can conveniently place one's head—or, perhaps rather
we should say, here is a particular kind of thinking-cap that we
can put on—and it stands as a quasi-technical method for producing quasi-technical results. Those results are calculated to be of
more immediate utility for the ordinary purpose of running the
machinery of the world than profounder disquisitions on the
fundamental problems of life and destiny.

So, instead of being narrators of a story in which one is never
supposed to know what is going to happen next, the historians (on
this alternative view of their function) may take the line that, at
any rate now, the past can never be other than it actually was. Accepting this as given, they may spread the whole of it like a map
before their eyes, and then set out upon the task of discovering all
possible kinds of correspondences and correlations between the
parts. The scientific procedures are capable of a peculiar beauty,
and I have no complaint to make against them, except that I
should like to see them extended to everything—I should like to
see falling in love, conversions to Roman Catholicism, national
aggression, atrocities in war, infant prodigies and the incidence of
the appreciation of Wordsworth's poetry, all subject to that en-
chanting technique of methodical collation which so often pro-

duces results far beyond anything we should ever have inferred in a course of arm-chair reflection. Those results are liable to be interesting even when they are in no sense absolute, as when my friend, the late Professor Schöfler of Cologne, showed how many of the figures in the Romantic movement had been products of the manse. If, after reading twenty detective-stories by a given author, I learn that, wherever the amateur detective goes— whether to stay in Devonshire with a friend, or to Florence for a honeymoon, or to Westmorland to see his mother, the Duchess— a murder occurs at his very side, I find it as irresistible as anybody else to infer that this amateur detective is a danger to society; and I agree with the stodgy official from Scotland Yard who said that, though executions for homicide were all very well, the real way to stop the murders would be to liquidate the amateur detective. It seems to have been established in my household (after events within the family had been correlated with information from out- side) that the strumming of an apparently innocuous sentimental piece by Tchaikowsky is always followed by news of a cata- strophic illness or death very close at hand. This example pro- vides an opportunity for noting the fact that the knowledge which is achieved by such a policy of methodical collation is defi- nitely the kind of knowledge which is significant for the purposes of action; for if I play the offending piece or pull the music out of the cupboard—if I even hum the tune in my bath—it is clear that I am being malevolently sadistic and that men of good feeling must combine to secure my suppression.

III

In the richest kind of historical writing, like that of Ranke and Acton, both of the two types of history which have been de- scribed will be found to be combined, so that there is interaction between them. On the one hand there is the narrative historian who in a certain sense is trying all the time not to know what is going to happen next. On the other hand there is the more scientific student who has traversed much of world-history with an eye for the correspondences and correlations, and an insight into historical processes. And here is one of the arguments for the importance of what we call the "general historian"—the man who, if he is describing a single revolution, writes with the added experience that comes from a great deal of pondering over all the revolutions of history. At this point the notion of personality in history meets the notion of process in history and the two form a new texture—they become fused in a higher synthesis. Furthermore, the story which the historian has to narrate must itself suffer a change when the new dimension has been added to it—when it is re-told after submission to the more scientific kind of analysis. And, above all, though the responsibility of the individual can never be eliminated, its location comes to be a less simple matter. Sometimes, even, the main burden of responsibility will fall in a different place.

One of the clearest examples I know of the redistribution and transposition of moral responsibility is provided by Mr. Gladstone in a kind of moral lecture which he delivered in writing to Queen Victoria and which I have lately seen quoted. It was in May 1869, at a moment when Gladstone was giving an important turn to England's Irish policy, and Queen Victoria, who was unsympathetic to this, was more concerned with the agrarian outrages that were taking place in Ireland, and with the need for repression and retributive action. Gladstone adopted the view that, human nature being what it is, certain conditions may be expected to produce dreadful consequences; that punitive measures against the obvious culprits would not meet the case, since the conditions needed to be changed; and that the evil was one which it might take generations to remove. He wrote:

> Your Majesty's advisers deeply regret the recurrence of agrarian outrages in Ireland; but they can feel no surprise at it. The movements of disease cannot be predicted with pre-

cision, either in individuals or in nations, and this class of
crime in Ireland partakes, much more largely than is common, of the nature of disease. Individual depravity has less to
do with it, evil tradition more. . . . The full fruits of the
work [the government] have undertaken, supposing their
judgement to be right, can only be reaped in the future . . .
and the patience of years, if not of generations, may be required in order to repair consequences which have come from
the perverseness of centuries.

Gladstone does not deny that the perpetrators of agrarian outrages have a responsibility which their priests or the police court
might bring home to them, but he sees a genealogy of guilt some
of which goes back to landlords and to Englishmen themselves.
The blame which he attaches to the obvious culprits is not measured merely by the magnitude of the consequences of the offence,
but is related also to the degree to which the offenders were provoked or were under special temptation. Indeed, while recognizing that every man has it in his power to make himself an exception to the rule, and some have a special power from heaven to
enable them to do so, we might well base a maxim on the thesis
that certain conditions will produce agrarian outrages, human nature being what it is, and that the evils are to be remedied by
altering the conditions. And, in fact, when that point is recognized
in the way that Gladstone recognized it, man can move a step further towards the elimination of agrarian outrages and towards the
control of his own history. By diagnosing and recognizing where
he is unfree, man may increase his power and steal a march on the
whole system of necessity. In this sense human beings still assert
their freedom; and the individual has the last word.

We have reached a point at which imagination and intellect
must combine; for even in history we might say that we can have
something which might well be called a "science," provided we
can conceive of a science that is to be handled with great flexibility. The application of what at least are quasi-scientific procedures
in the field of human studies produces in one of its aspects the

kind of laws which are only valid "other things being equal." Occasionally it seems to provide a norm from which some kind of departure may occur in any actual case. It may issue in the thesis: "human nature being what it is," such and such a piece of political high-handedness is calculated to provoke a population to desperate measures. Alternatively, it may lead to generalizations about large masses of people, and, though there may always be some exceptions, there are kinds of arguments in which such generalizations are valid, because the exceptions do not affect the case at issue. Apart from anything else, there are regions in which something like laws may be established, and yet these laws may be superseded by the mere fact that men have become conscious of them—the very realization that they exist, or that the tendency exists, is sufficient in fact to alter the state of the question. To take the crudest kind of example, the view that revolutions are subject to a process which makes them tend to move ever further to the left, has been contradicted in a number of historical instances; but, curiously enough, the contradictions really show that where men have been most conscious of the tendency they have best been able to check it, for they have known to take special measures to counteract it. Such an exception does not invalidate a general rule when what is in question is the kind of knowledge that is subordinate to action; for the original generalization is merely subsumed at a higher level in a wider one. The very case for these quasi-scientific procedures is that the knowledge we acquire from them can itself be used to some effect in the world; the possession of the knowledge is important precisely because it does alter the state of the question. If, therefore, they have a realization of the general rule, men are able to go one degree further towards the control of their history and destiny; and if the "science" that is acquired is only a flexible one, it is of the kind which we might call "wisdom" or "experience." Even at the end of this argument, men as individuals come out on top—by learning the processes of history they can go one step further in the attempt to gain direction over them. The individual emerges as being more highly conditioned by history than *prima facie* we had imagined; but the very fact that he has come to understand this, and to be more scientific in his attitude towards it, means that he is able to steal a further march on events, able to conquer a wider area for the exercise of his will. The Napoleons and the Lenins of the world have always been highly conscious of this fact; but there is no reason why this kind of "science"—this conscious exploitation of the process of history—should serve merely as the ally of the tyrant. Those who most believe in the importance of personality in history, therefore, are the very ones who ought to welcome this other aspect of the story. They are the people the least likely to suffer damage from it or to make anti-humanist inferences from it. They have to take care not to leave entirely to the enemy a weapon of such formidable power. They can see greater splendour in Creation and find more exhilaration in it just because both these aspects of the truth of history have such genuine validity.

Lord Acton (after first fighting Buckle) came over to the view that the Positivists in their search for laws would do better for history than the Bossuets and those people who write their narrative so as to show on which side God is fighting. Let us welcome this treatment of history, therefore, and give due honour even to the Positivists lest they become too disheartened in these days. What is important is that the other kind of history should preside—that indeed the Humanities should preside over all the Sciences, which are only the servants of the servants of God. And I suspect that the Humanities are never killed by the Sciences, but tend to commit suicide because they cannot bear to live any longer with their own inferiority complex; when in reality they ought to slide into the presidential chair as though they owned it—especially as what chair seems otherwise to be empty. What we have to do is to capture all that these sciences have to say, and use it for our purposes; like those Christians of an earlier age who, whenever they saw anything good, boldly claimed it for themselves—blithely laying hold on anything they liked even from pagan Greece, and (no matter whether it were the Stoics or Aristotle) insisting with admirable effrontery that it was their own.

For all these reasons all those people who feel inclined to stress the importance of personality in history, seem to me to be mistaken if they fear the extension of scientific procedures or fail to capture all the resources of science for themselves.

3
Christianity and Politics

The interpolation of the religious factor into politics has always been one of the disturbing features of general history; and it would appear that no special providence has particularly protected the Christian religion against the possible dangers. Such an intervention has often meant an insistence upon policies that would safeguard the power or the vested interests of established ecclesiastical systems. Sometimes it has meant an attempt to impose upon non-Christians certain obligations that had no meaning save for those who accepted the faith. But also it has often involved the insertion into politics—even into a war—of an element of fanaticism which itself can create the most serious problems for a benevolent statesman. There is a kind of mentality that will not be satisfied until Napoleon, besides being a tyrant, is shown to be the very Beast envisaged by the Book of the Revelation.

The pulpit played a considerable part in the promulgation in England of the unfortunate view that the First World War was a "war for righteousness." English churchmen, when they condemned the German churchmen for accepting the propaganda of their own government in respect of the origins of that war, were only too unaware that they were imprisoned in a similar way themselves. In one war after another Christians may give most inspiring examples of solicitude for the suffering, and care for the wounded; but episcopal declarations on the issue of a conflict, by mixing religion into the matter and claiming God for the home cause, can add to the fever, and may appear unfortunate after a lapse of time. There are periods of history—times of religious war and persecution—when, precisely because of their zeal for a certain kind of righteousness, the ecclesiastics have wanted to continue the killing after worldly-minded politicians had felt that there had been too much of it. There are periods when, so to speak, a secular ethic has had to be called in to rectify the aberrations of what claimed to be a supernatural ethic. Along with that great defender of the Church of England, Lord Clarendon, we must sadly accept the fact that the affairs of the world would be worse if we were governed by ecclesiastical statesmen. The Christian who feels that his religion has anything to contribute to the politics of today must realize that the outsider is going to be very cautious of him. He had better disguise his message as common sense.

I

There are certain great themes on which the student of history ought ceaselessly to meditate. One of these is the role of religion in the history of the human race. The subject provides a tremendous background to any discussion the Christian may have with his conscience about the relations between his beliefs and the problems of the contemporary world. A glimpse of this imposing background, or at least a sense of its existence, would be useful to those who feel that their faith enables them to make a contribution in the realm of current affairs. It would be equally relevant for those who see danger in the intrusion of the religious factor into politics.

The purely secular historian might be tempted to say that religion performs its most magnificent service when society is comparatively primitive and a culture is in its formative stages. In Europe, for example, a religion, which regarded its truths as absolute and felt itself to be "universal," presided for something like a thousand years over the development of civilization. It would appear that a faith which has so established itself may give a powerful lift to people who are ready to emerge from barbarism. The necessary prerequisite is that the religion shall acquire a unique ascendancy over a considerable stretch of territory. Genuine "universality" is never actually achieved, however; and what we find in fact is that there will be one vast area in which all men will naturally believe in the Koran, while in another area there may well exist a similar, almost hereditary, almost inescapable, belief in the Bible. It is even true that in these two cases the ascendancy could hardly have been achieved, the virtual unanimity hardly procured in the first place, without the use of authoritarian methods not radically different from those which produced the spread of communism in the twentieth century. But in days when the herd-spirit was strong, authoritarianism was appropriate to the needs of society; and, if we are to judge from the plight of those regions that escaped the Church in medieval Europe, the alternative was likely to have been the prolongation of barbarism. Lord Acton, more than any other historian, hated persecution; but he pointed out in his manuscript notes that if, at a certain stage in the story, the Christians had been too squeamish to use force, the result would only have been the conquest of Europe by Islam. And, indeed, this would have been the next best thing for Europe.

So far as Christianity is concerned, it adapted itself to the kind of world in which it found itself, and it assumed forms that would hardly have been predicted from the mere reading of the New Testament. In the mundane sphere, it performed the kind of functions that other religions—even those called "pagan"—so often performed elsewhere; and in this respect the closest analogy would be the part played by Islam in the Arabian world. It acted as the cement of society and the bond of the tribe. Its powerful influence over what might be called opinion was used in general to support the secular rulers. It was in a position to demand a great deal for itself in return. And its representatives were the intelligentsia of the time. There were certain aspects of the faith itself, the doctrinal teaching, that were calculated to influence the ordering of society, the development of culture and, sometimes, the actual turn of political events. In the Middle Ages it was significant, for example, that Islam laid a peculiar stress on the Law. In Europe things worked differently on many occasions because, in Christianity, the corresponding emphasis was rather on Doctrine. At the same time there were similarities between the two religions and these would reproduce themselves in parallel cultural developments. In the case of both Christianity and Islam it would seem to have been the influence of the Old Testament which operated to produce an interest in the past and a regard for history not shared by some other religions and civilizations.

The services of religion to the cause of mundane progress were great, particularly as there were some religious valuations that had a mundane reference and affected the conduct of life. By a process of elision these came gradually to be turned into secular values, and, in that capacity, they ultimately achieved a sort of autonomy. They endured after the spiritual element had evaporated out of them, surviving even in men who had lost the faith out of which the ideas had been born. The origin of these is often forgotten, but, even in their secular form—as "Western values"—they are still sometimes described as "Christian," though in this there may be some propagandist intention. We can hardly measure what the modern doctrine of individualism must owe to the Christian belief that men are spiritual beings, born for eternity, and having a value incommensurate with that of anything else in the created universe. Because all human beings were regarded as equal in the eyes of God some men were ready to assert the egalitarian principle in human society, a thing not always so plausible as it is to us today.

The emergence of conflict—spiritual versus temporal—in the Middle Ages produced a breach in the system, a gateway for modern freedom. There might be many principalities or kingdoms in a Christendom that had the dimensions of a continent, but the presiding religious authority was bound to deplore dissensions within the system, and it tended to mitigate these. It attempted to regulate the relations between peoples and to bridle the horrors of war. By the time the modern states had emerged it had assembled much of the material for the first projects of "international law." Religious authority sought furthermore to procure cooperation between Christian governments, especially when there was a threat that came from outside the system. Conflict was likely to be more grim when Christianity and Islam confronted one another and the war between empires and cultures was a war against the Infidel.

In all this we have something the New Testament did not cater to: the actualization of the beautiful dream of almost a whole continent standing as a religious society, the people virtually unanimous in the faith. And Christianity—like modern communism, no doubt—was able to ensure the effectiveness of its influence and the endurance of its system by its command over education. All discussions of the relations between the Church and the world depend on the answer to the question whether such a unanimity in the faith is feasible, whether such a dream of a "Christian Society" is legitimate, save in those younger days of the world when the herd-spirit was strong and authoritarian government inescapable. Even today, the mere suspicion that such a religio-political system might return in the case of a revival of religion (though nobody might be thinking of such a thing at the moment) is sufficient to freeze the heart of anyone outside the faith. And perhaps because Christians cannot imagine such a thing occurring, or see any great harm in it if it did occur—cannot picture any harm that they themselves would do if they had power in the world and do not remember that, when the Church is allied with power, this itself is sufficient to bring the wrong people to the top—because of all this, men are perhaps right to be apprehensive. The feelings of these latter are intensified because, almost immediately behind them, more directly within their vision, are the centuries in which the whole system was breaking down. And it is in the breakdown of the system that the latent evils of the "Christian Society"—this particular attempt to mix Christianity with the world—become so appallingly demonstrable. What is beautiful enough when it is the result of a widespread unanimity, shows itself as tyranny when it fights for its own prolongation after the unanimity has gone. Christian influence is still suspect in many quarters because the resentment is so strong, because the last memories of what was felt to be tyranny are still so close. This helps to account for a strain of bitterness in the "Humanists" in England. It blocks people's minds against the Christian message before they have heard what the Gospel is really about. One may look at the Christian religion and ask: "Can such a faith offend?" But innocence is not enough; and prevailing modes of thought do not allow people to see how much of what has happened since the Renaissance has been the result of tragic failures in ecclesiastical leadership.

In spite of all the ecclesiastical abuses in the background, it is possible that the Reformation occurred partly because the medieval Church had done its work so well. If the barbarian peoples had been brought to Christianity by a kind of mass-conversion and by authoritarian means, the subsequent centuries saw what was really a vast missionary enterprise that had to be conducted in detail—a process in which men as individuals were induced to bring the important aspects of religion home to themselves. Perhaps the really significant development on the eve of the Reformation was the activity of laymen in the Church; it was almost, on occasion, a resort to self-help on their part. And it is a question whether the Reformation could have succeeded but for the preaching—the conversion of laymen in the towns, for example—and whether even this would have been so effective had the Catholic priesthood not recently been so inadequate in that matter. The Reformation would not have been so radical, or the religious breach so irremediable, if the new preaching had not involved new doctrine—doctrine which, even in its aberrations, was particularly born from actual experience and the problem of its interpretation, and particularly calculated, therefore, to answer to the needs of men in general.

It might be argued that Providence did not fail the Church in respect to one matter: the light did not depart, the Gospel was preached, and pious men in every age were able to deepen their spiritual life. In thousands and thousands of villages and towns, there would be clergy who, week in and week out, year after year and century after century, called the attention of men to the fundamental principles of Christianity. All this slips through the net of the ordinary historian, but the total sum of it must have been important for our culture, perhaps as important as the activity of politicians. Also it must not be forgotten that, in the period after Luther, the religion still found expression in great literature.

So far as concerns the outer framework of the system, however, the Reformation is chiefly important because it established religion on a national basis. Lutheranism and Calvinism—and Catholicism, perforce—endeavored to establish themselves through alliance with governments. None was able to achieve the degree of "universality" that it claimed. Henceforward each separate nation was its own "Christian Society."

Because each claimed to be the true faith and held any perversion of this to be the work of the devil, because nobody could imagine that two different versions of Christianity could be true, the wars between nations turned into religious wars, and this added to their fanaticism and cruelty, lessening the possibility of compromise. In a sense all parties were fighting for the medieval ideal of the "Christian Society," each claiming it for itself and seeking a unique predominance; but, that the world had passed beyond this stage is demonstrated by the fact that so many of the

countries of Europe were unable to eliminate their own religious
minorities.

Because the "Christian Society" now coincided with the nation,
the authoritarianism in the system was more localized—the tyranny
closer at hand—than in former days. And, now, each of the rival
churches was more closely tied than before to the existing national
regime. The cause of Christianity itself would be linked to the
very fortunes of a regime, or to a certain theory of the monarchy.
Established churches became defenders of the status quo, holding
to the doctrine of the Divine Right of Kings (holding to it as almost an article of the faith), for example, long after all reasonableness had departed from it. In other ways, the survival of a
too materialistic conception of religion had tended to tie Christianity almost irrevocably to the things of the world. It was even
feared that the whole fate of the Christian faith depended on the
maintenance of the Aristotelian picture and theory of the cosmos.
For these reasons, the established churches in the modern centuries tended to resist the coming of more liberal regimes, and the
advance of the natural sciences, although the liberalism had been
born out of religious principles, and many of the scientists of the
crucial period were only conscious of working to the glory of
God. For one century after another, therefore, official churches
stood as though they had their backs to the wall, but presented
unnecessarily, until our time, the spectacle of bodies constantly
conscious of making an intellectual retreat. Even when Church
and State became separated, there remained the modern intensified
nationalism which constricted Christians. Extreme examples of
this appeared in Roman Catholicism, in spite of its international
system.

We are still doing less than justice to the case, however, unless
we separate the operation of Christian principles in society from
the political or mundane activity of Christians themselves, and
particularly of established ecclesiastical systems. Acton liked to
show that the principles of "Whiggism" went back to the Middle

Ages, and the seeds of modern progress, it must be remembered, emerge first of all in a religious context. What is here in question is the way in which the principles of Christianity (even certain of them that were intended to have an exclusively spiritual reference, and only by a kind of elision came to be transferred into the realm of human relations) came to work in society by a process that can almost be described as chemical: in a sense they lost themselves by combining with something else to produce a new thing. The most apt simile for a process so barely perceptible is that of the leaven which gradually works to leaven the whole lump.

These principles had their effect on society even while Christian leaders and ecclesiastical systems were rejecting their consequences; and sometimes they provided the very criteria by which, at a later date still, ecclesiastical systems themselves were judged. In this respect the mundane role of nonconformity has a great historical importance; for the blending of Church and State in the established Christian society put the religious nonconformist into a posture of hostility to the whole combined system—hostility both in matters of politics and in matters of religion. He not only applied the principles of universal Christianity in his criticism of the existing regime, but, so long as he did not have the existing regime to defend, he was more open to developments in thought, developments even in natural science. At the next stage in the story, the *philosophes* of eighteenth-century France are often "lapsed Christians," fighting sometimes for the things for which nonconformists fought. Their breach with the Church seemed to free them from the conventionalities that had constricted the application of the principle of Christian charity, and they made use of that principle even in their attack on ecclesiastical systems. Sometimes, since then, it has been the atheist who has been ahead of those systems in promoting causes now regarded as "Christian." Sometimes, therefore, in the mundane sphere, the principles of the religion have worked better than its agents. One would conclude from the whole story that it is a fine and fertile thing for Christians to cut themselves away on occasion from tradition and from the conventionalities, returning with utter fearlessness to the fountain of their original principles.

II

Although Christians may feel that they have an easy game when a high ecclesiastic has the ear of a powerful king, the effects of this conjunction are liable to be a disappointment, the world being what it is, and human nature being what it is. Cardinal Richelieu, though in many respects he has had a bad reputation among historians, does seem to have meditated with some earnestness on the relation between his religion and the conduct of public affairs. It would seem to be this which led him to produce the thesis that, even in time of actual war, the work of diplomacy should never be suspended. He so far lived up to the maxim as to be twitted with having started negotiating for the end of a war before hostilities had actually begun.

What we are concerned with, however, is a very different situation: a world of democracies in which every man has his responsibility for policy, and, on crucial occasions, the voice of every single man has as much weight as that of every other. The Christian has his own decisions to take as one of a multitude of co-rulers. But if he feels that he has anything further to contribute, he, like anybody else, can benefit from the right that any minority possesses to enlarge its forces, indeed to become the majority, provided it uses the method of persuasion. The question arises: Will he have anything particular to contribute, anything which, for example, would parallel Richelieu's maxim about the continued functioning of diplomacy in time of war? Can he legitimately differ—does he actually differ in any case—from the rest of modern secularized society?

We should probably be rendering ourselves open to a hundred

fallacies if we were seriously to label any civilization as "Christian"; and we might be inaccurate in certain ways—or we might do political harm—if we insisted that this Western civilization of ours is to be described as even "lapsed Christian." Christian influences have certainly soaked into its very fabric during the course of centuries; and if we envisage our world as totally secularized, we still might have to say that Christianity, or the Christian view of the human drama, has left its mark on the secularity of the present day. A similarly secularized Islamic culture, or a Chinese civilization, however nonreligious, would differ from ours, probably in a number of important ways, but certainly in a thousand subtle ways, which it would need the whole of history to elucidate. And when either of these is thoroughly overlaid with modern Western influences, we do not know what deep differences between a secularized East and a secularized West might not push through the crust, revealing the inescapability of the past in both cases. Still, the churchman today will hardly regard many of the obvious prejudices and assumptions of modern society as Christian. Where he does in fact agree with them he may differ in the distribution of the emphases. Where he feels them even to be Christian, he may go further in the application of them, either for that very reason, or because he has chosen to put himself closer to their original source. Alternatively his position may not allow him to overlook certain things which nobody disagrees with, but which the world in general finds it easy to forget. Apart from anything else, his connection with a historic belief may give him a sympathy (or a certain sense) for things that are the heritage of long-term human experience. If he has cleared his mind of all the vested interests associated with "historical Christianity," we might expect him to have a general attitude of his own toward the question of the way in which politics ought to be conducted.

It is sometimes the advantage of those who believe in God that their faith at least prevents their ascribing divinity to anything else—to individuals or great collectivities, to mundane systems or abstract nouns. Whereas for thousands of years political history almost invariably meant the history of one's own region, nation or empire or country or city, it always tended to be religion (or a quasi-religion) that encouraged "universal history" and envisaged the destiny of the whole human race. The doctrine of man's universal sin—and, more particularly, the teaching that one should look at one's own sins first—is calculated to be a serious check on many evils and mistakes in politics. At least it operates as a caution against that self-righteousness which so bedevils public controversies and embitters wars—the sin most disastrous of all in its political consequences, and sometimes the last remaining fault of the noble pagan, the most terrifying characteristic of some of the more virtuous of our modern youth. The unforgivingness that makes so many of our national or international problems so difficult may have been a fault of high and low in the Church sometimes, but it cannot be imputed to Christian principles.

It would be a good thing if men would recognize that, in the case of many of the world's conflicts, the struggle is not between right and wrong, but between one half-right that is too wilful and another half-right that is too proud. Then again, it might be a good thing if men, measuring the forces in the world, could be made to ask themselves in extreme cases whether, unless they change the direction of their efforts, they are not fighting against Providence. Leaders of the Church—upholders of fanatical causes— may have been worse sinners than anybody else about the matter; but the better side of the Christian tradition would hold that even a good thing may turn into ashes in our hands, or may be transformed into evil, if it is achieved only by too great an exercise of power. And at least the members of the Church that survived the martyrdoms will be aware of the limits of what can be achieved by power, and will know the danger of *hybris* in this connection.

It is not clear that the practice of politics would be damaged by constant insistence on respect for personality, respect even for the other man's will, when that man's will is different from our own.

We may be making things too easy for ourselves if we pretend to wipe out the members of another party as fools or rogues or mere agents of vested interests. The preaching of mercy and humility and charity in countless churches through almost countless years must have had its effect on the Western idea of personality. Constant contact with such homely homily would do no hurt to many of the practitioners of politics. One might add that it would be inconsistent with Christian principles to seek to prevent a given evil by means which themselves would have the effect of installing in the world an evil greater still, though of a different sort. And the notion that, by such a measure, a certain Justice, high in the cosmos, would be vindicated (vindicated as against one particular breed of sinner, while leaving so much injustice still rampant in the world) would be like a reversion to pagan superstition.

Difficulties about these various points have tended to arise out of misunderstandings concerning the nature of the Christian ethic. The New Testament is otherworldly in a certain sense, for the believer in Christ is enjoined to put his heart and his treasure in heaven. But the religion always had one foot on the hard earth, and if it stressed the divinity of Christ, it also insisted on His complete humanity. And so, keeping its link with the soil, it helped to give Western culture its high valuation of mundane history. It can produce a kind of worldly-mindedness cleansed from the myths that do so easily beset us.

In a sense, Christianity is in reality less an ethic than an insistence on a way of life. It has one law, the law of Love or Charity, and this—a definite and demanding thing, though so infinitely flexible—is the source of whatever the Christian may possess in the way of rules and regulations. There is no differentiation within it save that, as Maritain once said, if, in a given situation, you can think of an act more charitable (all things considered) than what you are doing, you must devote yourself to that; and, no doubt, if you are unable to think of anything more charitable, this itself is because of your own limitations and defects. Precise rules, precise working-out of the effects of the basic principle, have been put forward throughout the ages; and perhaps it was not always so unfortunate as it might seem that many of the rules envisaged the maintenance of an existing social order or the support of an established regime. One of the misfortunes of the historical Church is its inflexibility sometimes in regard to these prescriptions, a mistake which so often set ecclesiastical systems against what Christians themselves would now regard as the light. But in any case, one of the strongest traditions of Christianity is the fact that, in spite of its aspect as an otherworldly religion, it attached importance to the conduct of men in the world, imputed responsibility to them, and inculcated a profound concern for suffering anywhere.

A big difference between the West and the East until not very long ago was the concern the young Westerner (even the youth of undergraduate age) might have for men at the other end of the globe—the feeling which possessed him that he had a certain responsibility toward them. It is in the history of the Church that one can see something of the emergence of this care, whether for the souls or the bodies of people sometimes far removed. The same law of Charity enjoins a serious solicitude for live men as they actually exist, not merely a postponed and doctrinaire affection for hypothetical people in a remote and highly questionable future. It means no specialized love, no concentration on politically slanted cases, but a vision which includes the North Vietnamese and the South Vietnamese, the blacks and the whites alike, and the vicissitudes of civilization itself. There is no resort to an easy political solution by closing one's mind to every aspect of a problem save one. Statesmanship is judged by the catholicity with which it gathers up all the varying factors in the case, and by the balance it strikes between them.

With these things in mind, you look at the world as it stands in its stark reality today, and you do not brood on good things that have been lost, you do not have obsessions about the recovery of what once was prized. You simply ask: How can this existing world be made better at the next turn in the story? So you do not sacrifice a generation of real live men for the sake of a hypothetical good in the distant future. In the case of war you do not allow yourself to be dominated by the question: Who began it? You do not build fantastic card-castles for the future, assuming that your successors will be virtuous and that your grandchildren's purposes will be the same as your own; such skyscrapers are going to collapse in any case if a single card is put slightly out of place. The Christian knows the defects of human nature as one who is only too conscious of the share he has in them. And this ought to act as a further safeguard against one cruel aberration: political utopianism.

III

But it remains true that religion is a dangerous substance. There are features in "historical Christianity" which show the perils of a religion that has moved too far from the fountain of its original principles. It is a curious thing that the New Testament faith has not always—as it ought to do—saved men from the crime of treating politics fanatically, as though it were religion. It is still possible at the present day to see signs of a thing that might be a pitfall for unthinking Christians: a kind of inverted utopianism in the reaction against communism. At worst this might lead to the greatest catastrophe of all, namely, a war or a rivalry staged as a conflict between Christianity and communism. At best it is apt to introduce a recalcitrancy and unreasonableness that are calculated to defeat all the efforts of the diplomats.

In this connection there are some important points of history the churchman ought never to be allowed to forget. If there is an orthodox teaching about the Crucifixion, it is that the responsibility falls on man's universal sin. Christians are allowed to treat

the event only in a certain way. They simply have to say to themselves: "We did it." Nothing could have been a greater perversion
than to turn the whole affair into a crime imputable to the Jews
as a race or a nation. It would be impossible to measure the part
which this kind of thing played in the development of European
anti-Semitism. But something of the same argument has its application when Christians make a particular attack on the intolerance
and the authoritarianism of the communist system. What can they
decently say, except "We did it"? What can they properly infer,
except that there is something wrong with human nature?

The thing that makes one hold one's breath is the degree to
which Communist Russia mimicked the system of "historical
Christianity" even in the cruelty of some of its methods. And if
we say that this latter was three or four centuries ago, is there
not something too facile about this argument? We can withhold
our anger, and feel that perhaps, under the same conditions, we
would have stood a good chance of behaving in the same way.
We can open our eyes and learn to understand the thinking of
people not like-minded with ourselves. It is even true that the
whole device of reducing a great number of the inhabitants of
a country to the status of "second-class citizens" was developed
in Christian Europe, and achieved considerable importance in the
modern Protestant world. If the passage of time or the advance
of civilization or the release from emergency conditions enabled
persecution to be mitigated, and opened the way for the establishment of modern liberty, why should not something of this sort
happen over again as the twentieth century proceeds? In any case,
nobody must argue that the Christian dispensation as such made
these types of tyranny impossible. The myth—so apparent in our
time—that there is only one last enemy of civilization to be fought,
and this one the greatest of all (so that its defeat will make the
world "safe for democracy"), can be made to wear the mask of
religion, but it is opposed to the very essentials of Christian doctrine.

All Christians, all historians, must be a little sorry for statesmen, who confront problems that are overfacing and always have fewer options at their disposal than the outsider realizes. In the last quarter of a century the governments of the world have steered us through crises that were far harder riddles for the diplomats than that of July 1914—problems not actually produced by the people who had to solve them. But even Bismarck, who might seem a man of mighty volitions, repeatedly made it clear that the statesman ought to allow the processes of time to do half his work for him. And perhaps those are of greatest service who do not try to assert a sovereign will, but take the line that they must cooperate with Providence.

4
The Conflict Between Right and Wrong
in History

I

Many of the people who nowadays belong to an older generation will remember having been brought up on a kind of history which tends to see the whole human drama, the story of all the centuries, as a straight conflict of right *versus* wrong. On this view, when one vast system confronts another, when power-organisations get into conflict with one another, this is a neat case of the good men fighting the villains—*our* side, of course, being the virtuous ones—our side not only struggling for the superior kind of system, but better as individuals, finer men, even braver on the field of battle.

For those of us who half a century ago were brought up in the world of "nonconformity" in England, this meant that so far as bygone ages were concerned, we were on the side of the Protestants against the Catholics, the Puritans against Charles I, the nonconformists against the Church of England, the Whigs against the Tories, and the English against the French. In the days of my own childhood, it was still the English against the French, these latter being the traditional enemy. I can remember even now the schoolbook which said that the English owed all their freedom to their kinship with the Germans, for liberty went back to the Teutons in their primeval forests. The Reformation, the emancipation of religion, came from Martin Luther, and Germany in any case had long enjoyed federal government, state rights and even free, independent, self-governing cities, like Hamburg. The antithesis to all this was to be found in the Latin countries. I still remember how it was all spelt out: Italy stood for the Papacy, Spain had had the Inquisition, while France, twice over, if you please, had chosen to live under Napoleonic dictatorships, an evil which, in my young days, had as yet had no parallel in other countries.

By the grace of God, those people who in the great conflict supported the wrong system—the Catholics, the Tories, the Frenchmen for example—were personally criminal too. They persecuted, they committed atrocities. Once in a while there might be a man on the enemy side whose virtues could not quite be overlooked, but you would mention these only in parenthesis. There came a time when we made rather a joke of this. You would just interpolate, "though respectable in his private life." Nevertheless this was a minor consideration compared with the sin of being on the wrong side in Church or State.

On the top of everything else, the enemy, the other party, when the conflict was over, would poison and pervert the whole history of the affair. The truth was that, when we dealt with the past, our side did very much what the other side was doing. We produced the type of history that I call commemorative—the sort that the victorious party writes after the war is over, when you count your trophies and gloat over the defeated and fight the old battles over again. I can still picture very vividly the place and occasion when I discovered that the Left Wing in England could tell untruths as well as the Right; that the people on our side could distort history too. I was suddenly stunned with the realisation that I should have to reshuffle everything in my mind. It hardly seemed possible to find bedrock anywhere if both sides were producing partisan histories, and both sides were occasionally rather unscrupulous about the matter as well. That was one of the reasons why I became very grateful, when I went further into the study of the past, to find that scholarship at its highest levels was adopting a very different form, much of it in England actually changing its character, producing new ways of setting out the story, new ways of staging the drama of history. These latter, incidentally, worked with a more healing effect, a greater tendency to promote the reconciliation of man with man.

When I first began to lecture in Cambridge, I remember feeling very lucky because, at the end of the nineteenth century and in the early decades of the twentieth, foreign scholarship, after a considerable period of development, had come at last to a thorough re-structuring of early modern European History. I was able to pass this on to my students as the latest thing. It had not yet been quite filtered down into our educational system. On big topics like the French Wars of Religion, the Revolt of the Netherlands and the reign of Philip II of Spain, researchers had been collecting the manuscript materials for decades but had gone on poking the new evidence into the ancient framework of narrative, still mounting the story in the way it had been told by the original combatants. They went on squeezing the new source material into this antiquated structure until all the lines of the old pattern were bulging and the thing had ceased to be plausible. But now they had come to the point when the only thing they could do was to break the whole story down again, and reconstruct it entirely, staging the drama in a different way.

In a sense it was as though the historian had moved to a higher altitude, from which he could see both parties to the conflict at once, instead of just glorying in his onesidedness. I felt that it was as though, now, you suddenly saw everything stereoscopically—a new dimension of the picture seemed to stand out, unexpected things being brought to the forefront. During the same period, a similar change had been taking place in the study of English history, an undermining of what we call the "Whig interpretation." Some germinal things had been occurring in the 1890's, these themselves due partly to foreign historians who were outside our party politics, and who, precisely because they were strangers, managed to escape many of our local prejudices. In all these changes that were taking place in historical scholarship, the thing most affected, of course, was the way in which you had to envisage these colossal conflicts that take place between great collectivities, between human beings in the mass.

II

It may be useful, therefore, if we can persuade ourselves to reflect a little on the large question of human conflict, which stands out as a theme in itself, one of the great problems of human destiny. The straight fight of right against wrong does take place at times of course, but when the controversy lies between systems and super-organisations, a lot of threads come to be twisted together and the issues tend to become more complicated. Often in fact one finds that here is the sort of warfare in which reasonable and moderately virtuous men are taking part on both sides.

Of course there is evil in the world. It happens only to be more insidious than people often think, and we are all involved in it more or less, even the moderately virtuous. Part of the trouble is due to the fact that the moderately virtuous are by no means perfect, by no means free of vested interests for example, and there is a mixture of virtue and concealed egotism which is perhaps a greater cause of political conflict, a greater source of political problems, than anything else on this globe. At any period of history, these people who are trying to be virtuous, and who feel themselves virtuous, look with dismay on the world in general, and, in the face of this complex situation they tend to feel that they are hamstrung. One can turn to any period of history, and the picture is very much the same. Mankind seems to be confronted by a terrible inheritance; men wonder whether their generation will just be able to scrape through and they feel that the

heavens will fall after they have passed off the stage. In 1904—a
period about which my generation was particularly jealous, because we looked back to it as a time of comparative ease and security—even then, I am interested to notice, G. K. Chesterton found
it necessary to console his contemporaries. He told them to cheer
up, for the world always had been on its last legs. And the mixed
and muddled society into which we are born, with the entanglement of evil and good and the botched-up compromises, is partly
the product of the desperateness of the situations in which our
predecessors had found themselves. For they too, we must remember, had often just scraped through by the skin of their
teeth. When I was young, we blamed our fathers and said that
we might have managed to rectify things if only our immediate
predecessors had not fouled the field. And our fathers deserved
what they got, in a way, for they had treated their parents in the
same manner. It is a story that goes back *ad infinitum*. There was
a time when my Roman Catholic friends used to tell me that it all
went back to the Reformation; nothing in the Western world
ever went right after that, they said. Some of us who were Protestants managed to trump that ace—we said that it all went back to
the fourth century A.D. when a Roman Emperor, Constantine,
was converted and Christianity came to be tied up with govern-
mental power and vested interests. We ought to have gone back
further still, for the people in the early civilisations beat us hollow
on this issue. They pointed to the Fall of Man. After that, the
world could never be the same again. That feeling of exasperation
which many people have in the face of present-day issues—that
resentment at the monstrous way in which good endeavours are
thwarted—is due to the deep-seated nature of evil, and particularly
social evil. Very often one thinks that one is attacking the source
of it when one is really attacking only the symptoms.

We can say then, like all our ancestors, that ours is a world already spoiled before any one of us was born. We feel ourselves
confronted by entrenched and rooted systems, terrible organisa-

tions—it might be proper to call them organisations of cupidity. And, though they have suffered considerable mutations in the course of centuries, so that they are not quite the same now as in the society of the feudal epoch, there is a sense in which it is these that have been defeating the human race for thousands of years. Fundamentally, these systems—including the social order itself— are elaborate organisations that cater to man's cupidities, and that is why they are always such an ugly affair in every country in the world. All the same, they are not the worst that human nature is capable of, and there is a sense in which, as a matter of fact, they do also regulate the egotism of men and limit their cupidities. That is the reason why the world puts up with them. Certainly for a moment one feels like sweeping them away. But then one comes to realise that after they have been destroyed there is no question of having the reign of the saints, for the egotism is now more unbridled than before, and the weak are still more at the mercy of the strong. In this connection, I personally was greatly affected by a strike which took place in Liverpool in 1919, a strike on the part of the police, a class of people who had never seemed to be doing very much except standing in the streets (unarmed of course) or directing the traffic. The anarchy that was produced by the strike caused a lot of reflexion at the time because either it meant that there was a colossal criminal class beneath the surface in Liverpool and this had hitherto been kept under by the almost imperceptible action of these apparently very tame policemen, *or* it meant that a vast number of people who normally led respect- able lives in that city, looted shops and ran to violence as soon as they saw their neighbours doing it with impunity. There is al- ways more potential evil in the world than the social order per- mits to come to the surface; and, in the case of any resort to force, the malevolent agencies in the world are often willing to go fur- ther in violence and ruthlessness than the well-meaning people, when it comes to the point. At the same time, living under mas- sive systems and organisations human beings may need a jolt sometimes. They are in danger of just settling down under the existing system, doing their thinking within the system, under its ethos, or in terms of it, forgetting to what a degree it is a mixture of good and evil, an immoral as well as a moral order. They need to be jolted at times, forced to jump outside themselves, and to re-examine the whole system under which they live.

III

It is because evil is so deeply entrenched, so embedded in the past, so hardened by the passage of time—because also, in the course of ages it entangles itself so inextricably with the good—that we might ask ourselves whether there is anything to be learned by looking at the matter historically. It might be useful if we could see how the conflict between good and evil really mounts itself, and what it begins to look like when one tries to envisage it as a long-term affair. What is it that emerges when scholarship has revised the original jingoistic portrayal of events and one attempts to see things stereoscopically?

The character of the issue can be illustrated by an example which is simple because related to a single individual. Lord Acton, though a Roman Catholic, was perhaps more hostile to persecution than any other historian, and would insist on calling it murder even when the man whom he regarded as responsible for it was a Pope. When he was Regius Professor of Modern History in Cambridge, even the Protestants sometimes wished that he would change the record; they said that he had the Inquisition on the brain. However, he happened to overhear the way some undergraduates were talking about a particular case, and he intervened, agreeing with them that the man they were discussing was the most intolerant of persecutors, but pointing out that he was also the person who, when his town was visited by the plague, stayed on after all his colleagues had deserted, and died looking after the victims. We might well wonder, therefore, whether the historian should be entirely remorseless in his overall assessment of even this terrible persecutor. Judgment is easy for the fanatic who sees only one thing at a time, and so bases his verdict on only half of the picture. The difficulty comes when one tries to hold everything in one's mind at once and tries to envisage the composite picture and to view the case stereoscopically. Where historical writing is concerned, much always depends on the question whether a given scholar or writer will really face and properly admit the discrepant fact, the thing that does not fit in with the facile case that he is trying to make. And hence, when we are judging the quality of an historian, we have to measure amongst other things the amplitude, the catholicity, with which he takes *all* considerations into account. Everything is too easy if one is allowed to fix attention on just one thing.

The issues become more complicated if we turn our attention to a wider range of human happening, and see system confronting counter-system. Here our attitude to the question is liable to affect the way in which we draw the main lines of general history. The case can be illustrated again most clearly perhaps by a further glance at the problem of religious persecution. In the sixteenth century, Protestantism and Catholicism, as two mutually exclusive systems, were ranged against one another in the cosmic way in which communism and western democracy were ranged against one another after the Second World War. Many of us were no doubt brought up to believe that, in the terrible "wars of religion" which resulted, the Protestants were fighting for toleration in religion, and that to them—to their very bellicosity—we owe our liberty in matters of religious faith. As time goes on, however, I think that many even of the people who today would identify themselves with the Reformation, would come to feel that the bellicosity of both Catholics and Protestants was unfortunate for the world at large, and in fact proved particularly harmful even for the cause of religion.

Furthermore, there are signs that religious toleration was coming on very nicely before the Reformation took place, in the period of the Renaissance, the age of Erasmus; and this cause was precisely the thing that suffered so much from the bitterness of those religious obsessions, religious fanaticism and religious wars that tore the fabric of Europe and produced so much bloodshed. Martin Luther was not really fighting for toleration but for right religion, which he wanted to establish everywhere. Indeed, his very first moves were to secure the abolition of certain things (certain very unfortunate things, I believe) which he thought the Catholic Church was tolerating too easily. Whenever the Protestants won the victory and captured the government, they established their own intolerant system. One type or sect of Protestantism was very quickly persecuting another type. Since the persecution came to be carried on by the various local governments, it could be said that the tyranny was now worse than before, because closer at hand for everybody. The Catholics may have persecuted more in point of numbers, but that was only because they had previously represented the Establishment all over Western Europe; they had held the *status quo* and happened to be the Church actually in power at the beginning of the story. In reality, once they got the power, once they captured the government, the Protestants tended to behave in a very similar way.

Lord Acton, that great expert on the history of persecution, said that in some respects the Protestant type of oppression was the worst. According to him, the Catholics held that when you burned a heretic you were killing the body so that the soul would live throughout eternity. The Protestants, however, sometimes supported their case with some of the most cruel of Old Testament texts, calling for the total destruction of "blasphemers." In other words, they leaned more to the view that those who promoted the Catholic mass were not fit to live. And of course the Protestants might be charged with persecuting on behalf of a religion which in a sense they had only just invented—they persecuted as blasphemy those very things which they themselves had treasured as Catholic doctrine until only a few years earlier. Furthermore, in certain countries where the Protestants demanded toleration for themselves, it was a powerful nobility that made the demand and it was the policy of the landowners to force all their tenants, all the peasantry under them, to be Protestants. In other words, it was claimed that the king was not to compel the nobles to be Catholics but the nobles were still to retain the power of forcing the Protestant religion on all their tenants. When there was one party or sect which at that time stood for toleration, because it promoted a non-dogmatic religion, it suffered an equal battering from both Catholics and Protestants.

This side of the picture ought to be remembered when charges are made against the Papacy and the Inquisition. In reality we have two religious parties which were alike in their moral indignation against one another on account of persecution and of atrocities in time of war. Both indeed believed that only the *right* religion was justified in persecuting, justified in killing blasphemers and punishing religious dissidents. Yet when we today look back at that age and examine it with this problem of religious liberty on our minds, we reach the horrifying result that the deep cause of the evils and the cruelties lay in the very things that Catholics and Protestants held in common. The trouble lay in the very notion that faith could be compulsorily imposed—one's own faith being absolutely right. From the point of view of religious liberty it was important for the world that neither party should win completely in the sixteenth century. The road to freedom would have been barred if either of the great divisions of Christianity could have imposed its system universally.

It is doubtful whether in the whole of European history there was a bigger long-term conflict of Right *versus* Wrong (a conflict more earnestly felt to be such by the people on both sides) or carried on with greater self-sacrifices, greater readiness for martyrdom, than the sixteenth-century struggle between religions. Yet so far as matters of conscience are concerned, it seems highly possible that liberty would have been achieved more quickly if no fanatical conflict had occurred. Liberty came to a considerable degree through the advance of civilisation in general. In other words, many of the good things of life have not been achieved so much by conflict—by one party defeating another party—as people often like to believe. Some are the product of a cooperative advance of the human race, the result of the slow growth of reasonableness among men.

IV

The more general problem of international conflict leads to considerations of a different kind. We over-simplify the issue and we see too direct a struggle between right and wrong if in the case of such a conflict we accept the views of either belligerent party. Certainly one statesman may be more ruthless in his ambitions than another, and there may well be people who would suffer any loss rather than go to war. These differences in personal attitude may affect the surface story—affect verdicts on individuals—without necessarily guiding us to the deeper causes of the trouble. A wider historical view and an analytical study of a world in which many states are trying to coexist soon confront us with a paradox and a problem which belong to the very structure of international relations. They show how much of the evils of war is due to the defects of human nature in general.

For a very long time it was held by theorists and practising diplomats that when a state reaches a position in which it knows that it can do what it likes with impunity, it will take to aggression, even though it may hitherto have been virtuous. It may even have been acquiring merit by resisting some previous aggressor, defending not only itself but also smaller states. For others as well as itself it may have sought to secure guarantees against attack, then further guarantees, and then more again. It may have ended by shifting the centre of gravity and, in a fundamental sense, changing places with its enemy, presenting the world with a new preponderant power, a menace from an unexpected quarter. Even if such a state is ruled by a virtuous monarch at the moment when it moves into the predominant position, this (said Fénelon) would not be likely to endure for very long. New pressures at court, new forces in the country at large, new points of tension in Europe would have the effect of bringing a different kind of leader to the top. In general it has been true that governments when they are in a certain predicament, and states when they achieve a certain degree of predominance, tend to fall into familiar patterns of behaviour, committing the very offences which they had condemned so radically when they were committed by other states in the selfsame situation and subject to the same pressures and temptations. The virtuous conduct of states can be curiously related to conditions, therefore. And in a remarkable way the morality of governments may depend even on the distribution of power.

In modern history, first Spain, then France, then Germany, then Russia became the predominant power in Europe. In the seventeenth century Britain was slow in realising that the France of Louis XIV had taken the place of Spain as the menace to the European continent. After the Napoleonic wars, Britain insisted on strengthening Prussia against France in the Rhineland, so contributing to make that state the leader of what was soon an overpowerful Germany. Before 1946 it was almost impossible for people to imagine that Russia, if she emerged as the predominant power, would be as much a menace to the national freedom of the Czechs as Germany had been when she held a similar predominance. More curious still, the United States had seemed peculiarly unwilling to assume the responsibilities of a Great Power, and after the Second World War Englishmen wondered whether she would ever awaken to the danger from Stalinist Russia. It was not easy to convince anybody that she, too, when she came to the top of the world, would soon be condemned as an aggressor, a warmonger and an "imperialist" power. Perhaps she felt it her duty to build up her armaments and struggle for strategic positions, since in all this she had the idea that she was arming Righteousness itself. Perhaps any power that comes to the top of the world will excite the jealousy of others, however well it behaves; and if these others take measures for their safety, it will be driven to counter-measures, its very predominance generating therefore new anxieties for it. But at best, even here, cupidity is operative. And indeed the thrust and the strain to secure one's survival, and the lust for actual expansion, have never been confined to the Great Powers. Over and over again, the smaller states of Europe have taken to aggression in the same way, when a power-situation has provided them with a local opportunity. There is always in fact a lot of potential aggressiveness in the world, and if at given moments it has concealed itself, lying latent, this is often because power is widely distributed over the map, producing the necessary checks and balances. From the eruptions that have lately taken place in various parts of the globe, one can judge the ugliness of the vol-canic material that lies not far below the surface. Terrible things might happen to the international world—there would probably be a Balkanisation of everything—if we even lacked a handful of Greater Powers to keep the rest in order. There is more latent evil in the world than our national and international systems usu-ally allow to come to the surface.

But there are other difficulties. Let us consider an imaginary situation. Let us suppose that America and Russia had emerged from the Second World War exactly equally balanced, and that they were absolutely level in point of virtue, and in the moral qualities of the statesman who conducted their affairs. Let us assume that the moral level is a reasonably high one as statesmen go—the leaders of both countries not saints of course (not com-peting with one another in self-renunciation) but moderately virtuous men, desirous that their countries should come to no harm. Let us say that each is anxious to avoid a war, but each, by the nature of the case, can never be sure that this is the mood of the other. Each side will be sufficiently human to be beset by the devils of fear and suspicion, each will be locked in its own system of self-righteousness. Neither of them knows, neither of them has the possibility of knowing, the real intentions of the other. This is the absolute predicament, the irreducible dilemma, the thing that always tends to make human nature come out lower than usual, worse in a way than it really is. The greatest war in history could be produced without the intervention of any great criminals who were out to do deliberate harm in the world. In the circumstances that I have described, disarmament would be just as difficult as we ourselves find it at the present day, each side feeling it urgently necessary not to let the other steal a march on it in the course of the negotiation. And war, if it came, through some odd conjunction or some insoluble deadlock would be a terribly bitter "war of righteousness" with each side being so conscious of its own good intentions and feeling so strongly that the other has compelled it to go to war.

The predicament here described may not ever exist in its absoluteness, in its purity, but it enters as a constituent part into the constant relations between nation-states. It is the fundamenal reason for the tremendous tensions that exist. It is the reason why nations feel that they must arm, some of them by no means happy at having to undertake the measure. In a sense conflict is embedded in the very structure of the situation, though the virtues or vices of this statesman or that will produce a certain variety, or provide a kind of *violin obbligato*. And all the time, though each state is so aware of its own fears and apprehensions, it always fails to enter properly into the counter-fear of the other party, indeed never quite believes that with the latter it is a case of genuine fear. Similarly, each state thinks that the other is wilfully withholding the guarantees that would enable it to have a sense of security. Neither quite realises that no country in the world can have absolute security, the kind of security in which we can just rest.

No state can have absolute security—free from risk, free from strain—except on terms which would make that state a menace to everyone else. And all this has nothing to do with the issue presented by Communism. It is a standing feature of a multi-state system.

V

Let us be quite clear. The problem of evil is a very formidable thing—terrible because there is so much of this evil that is potentially there, lying in wait for the opportunity, so to speak. History gives us glimpses sometimes of the appalling things that can happen if the whole order of things breaks down, and if, for example, it comes to appear that there is no government capable of bridling the criminals. As we have seen, the world puts up with the idea of the state, and the existing order, because the evil and the good are in fact so intertwined. Some people may set out to replace the present system with an ideal form of Communism, for example, but even if this is pure at the moment of its establishment, it too comes to be mixed up with countless cupidities, and what one produces will turn out to be something more like a Soviet Russia. Indeed the Christian Church, though it started from the opposite extreme, and preaches everything that is the reverse of egotistical aggressiveness, can itself degenerate as it comes under the grip of the same forces. It, too, has come to seem at times like a great organisation of man's cupidities. We put up with the state because at least it does begin to regulate and limit these human egotisms; it is better than having no system at all, nothing to defend the weak against the strong. Perhaps we can say that it helps to get the human race a little way out of the jungle. If the whole order collapsed, all men would soon be having to learn now to defend themselves against one another; they would be having to carry pistols all the time. The state provides the framework for an order in which we are prevented from committing murder by fear of the law, but the prohibition becomes a habit and turns into an inhibition. So long as normality reigns, and we do not fear a possible murderer around every corner, it does not occur to us even to want to murder anybody.

When history is examined more deeply it becomes more clear that the good things of the world, the real advances in society, are not so often the result of conflict as has been imagined, and, even where they are connected with the victory of some men over others, the conflict is not often (as some like to believe) so directly between right and wrong. Interactions are constantly taking place in society, and individuals are constantly making their little contributions or their little obstructions. One cannot avoid being impressed at the finish by the way in which progress comes through these interactions. How often it comes even as the collective achievement of the human race, the cumulative result of innumerable people just doing a local good in their own corner of the world.

This would be illustrated by the prime case of religious liberty, which, as we have seen, has been made to appear too much as the victory of a party, the whole of history being ranged into a conflict between Freedom and Tyranny. It is wrong to imagine that liberty came merely as the result of conflict or could ever have properly established itself in that manner. We find ourselves under the necessity of tracing a number of very varied lines of development, those affecting different aspects of life and converging to produce a kind of world in which anything but religious liberty comes to be unthinkable. In reality religious liberty could not be established until a lot of other things had been changed, some of them things which we might imagine had nothing to do with the case. It required, for example, new notions of religion, new notions of society, new concepts of the state, new advances in general education. As we have seen, it is not possible to be sure that religious liberty was not delayed rather than hastened by the fanaticism of the Wars of Religion. During the stable period of the eighteenth century the march of reason was strongly evident in Western Europe, in spite of the enormous vested interests in society. The nobility themselves were sometimes the patrons of the *philosophes*. And, in England, universal suffrage was launched as a national cause in the year 1780, before the French Revolution had broken out. It is a moot question whether the Reform Bill in England was hastened or delayed by the outbreak of that Revolution. Perhaps reason was getting men towards the goal more quickly than violence could achieve the object.

Many of us will have been brought up to believe that democratical and egalitarian principles were really fathered upon the world and thrown into the arena by religious dissidents from the time of the Reformation, the nonconformists in England, for example. I personally would still feel that this was the case, and would hold that the nonconformists played a significant part in history. It is necessary that we should not lose our sense of humour, however. Those contributions on their part were due to the fact that they were a persecuted and disfranchised body. They were postured in an attitude of hostility to the whole Establishment, not only to the Church but also to the State which had adopted one confession rather than another. They were angled for opposition, predisposed to the criticism of the existing order. They lacked charity, however, and they were less virtuous than they appeared, for what they really wanted was to establish their own religious tyranny. Puritan England revealed something of what they hoped to achieve if they became masters of the country. It was important to have these religious dissidents, but it was almost more important still that they should remain a minority in a given country. Their virtue and their very attachment to liberty depended in reality on the fact that they stood simply as an opposition.

If England was fortunate to have her nonconformists, she was probably fortunate also in having the Church of England to supply a counter-influence and even on occasion a brake. Something of the same is true in the case of the Whigs and the Tories. Each party was stimulated, moderated and improved through the existence of the other, the Tories sometimes carrying a piece of Whig legislation or the Whigs a piece of Tory legislation, neither of them generally seeking when it acquired power to repeal what the other had enacted. In a sense, therefore, they may be said to have collaborated to make the England that finally emerged.

VI

It seems often to be assumed that the world consists of a lot of ordinary people who are very nice, but that life, and the course of history, are spoiled by the rare occurrence of extraordinary criminals. In the case of the great human conflicts, we sometimes satisfy ourselves with a facile view which fastens all the responsibility on one of these offenders, incidentally providing a neat exoneration for ourselves. The guilt of the offenders is magnified because evil is regarded as a kind of *deus ex machina*, and they themselves, being so exceptional, are seen as hostile interlopers, not really at one with the human race in general. It has to be admitted that the exceptional criminal does exist, and the point becomes obvious enough when, for example, we confront a Napoleon or a Hitler. But in the case of the great conflicts between nations or organised systems there is an anterior clash of purposes and interests which involves a standing issue of a fundamental kind. Sometimes the man who has come to stand out as the principal culprit should be regarded as only a supplementary factor in the case. He may have been exasperated by the deadlock, chosing therefore to cut the Gordian knot. Some will hold that in July 1914 this was roughly the position of the German Kaiser, Wilhelm II. Even Hitler comes upon the scene at one of these crises of exasperation, and there is a sense in which the British and their allies by offences of a less dramatic sort, helped to produce the predicament in Germany which gave him his opportunity. In fact, England at one time ridiculed this Hitler, regarding him only as we later came to regard Oswald Mosley. I remember a liberal-minded historian saying in great anger, in 1925, that unless Britain and her allies revised their attitude to Germany, this ridiculous Hitler might well turn out to be very formidable to us after all. Nor would the personal idiosyncrasies of Hitler have come to matter so greatly to the world, and to carry so much terror, if, behind him, millions of Germans had not allowed themselves to become too exasperated, each having some share in the responsibility, therefore, though their wilfulness was less obviously criminal.

On this view, history comes to look more like life, more like real life as it appears to any of us in the everyday world. The key to everything, including perhaps even the emergence of the extraordinary criminal, lies in the mediocre desires, the intellectual confusions, and the wilful moods of the ordinary man. It is important that we should fix our eyes upon this latter, that is to say, on what we regard as the moderately virtuous man, the average man, the man in the street. It is he (in other words, it is any of us) who puts sand into the mechanism of the watch, baffles the men who try to conduct government, and makes history a seamy affair. If the world's troubles were due only to one or two misbegotten criminals—if wars were always just clear conflicts between right and wrong—the problems of the statesman would be fabulously simplified. It is just the admixture of good that helps to build evil up into a colossal power. Alternatively, a very little wilfulness, a slight streak of petty egotism, can turn a grand benevolent project into a monstrosity. It is the moderate cupidity of Everyman, his ordinary desire to advance perhaps a little further than his father, his hope to sell more goods, and even his fear, even just his dread of a decline in his standard of living, for example, which gets history so badly tangled up and secures that events shall start tying themselves into knots. And, in history,

what might be felt to have been little sins can have vastly disproportionate consequences.

Wars would hardly be likely to occur if all men were Christian
saints, competing with one another in nothing, perhaps, save self-
renunciation. But the ordinary cupidities of little men, multiplied
by millions, build themselves up into a tremendous pressure on
governments, and can drive a population to colossal exasperations,
resentments, angers and political deadlocks. When this has taken
place, all the forces in society will soon be conspiring to bring the
wrong man to the top. Lloyd George once said that in July 1914
a great war might have been prevented if somewhere in Europe
there had been a diplomat who was more than an intellectual
mediocrity. I do not quarrel with this, but there are other ways
out, and I feel at the same time that if in July 1914 human beings
in general had been morally only a trifle better, the marginal difference would have been sufficient to prevent disaster. Thomas
Carlyle went further than this. After asking who was responsible
for the horrors of the French Revolution, he concluded, every
man in France, every single person who had done less than his
duty.

Therefore we are not quite adequate to the occasion if we
divide the world into right and wrong, and particularly if in 1914
we condemn Austria for the offence that the British will be committing in 1956. This diagram of what happened suits too well the
amour propre of the winning party in the struggle and enables
history to be blown up into a sort of mythology. The real conflict
between good and evil cannot be simply between one set of men
and another. Perhaps the very heart of it is the struggle waged
continually inside every individual human being. Furthermore,
the essential good—that which is fundamental and enduring—is
not merely the result of the victory of one party or nation over
another. It is more like the effect of a chemical substance operating in society, a case of winning people over, or planting a new

idea that soaks itself into the fabric of society, or just securing the slow growth of reasonableness among men. The real advance of the good is not by the defeat and discomfiture of some vested interest which conflicts with our own. It comes as an example of the most interesting of all historical processes—the operation of the leaven that leavens the whole lump.

II

Christians and the Interpretation
of History

II

Inference and the Interpretation
of Theory

5

The Originality of the Old Testament

I

The Old Testament sometimes seems very ancient, but the earliest considerable body of historical literature that we possess was being produced through a period of a thousand years and more before that. It consisted of what we call "annals," written in the first person singular by the heads of great empires which had their centre in Egypt or Mesopotamia or Asia Minor. These monarchs, often year by year, would produce accounts—quite detailed accounts sometimes—of their military campaigns. It is clear from what they say that one of their objects in life was to put their own personal achievements on record—their building feats, their prowess in the hunt, but also their victories in war. They show no sign of having had any interest in the past, but, amongst other things, they betray a great anxiety about the reputation they would have after they were dead. They did not look behind them to previous generations, but instead they produced what we should call the history of their own times, in a way rather like Winston Churchill producing his account of his wars against Germany in the twentieth century.

After this, however, a great surprise occurs. There emerges

from nowhere a people passionately interested in the past, dominated by an historical memory. It is clear that this is due to the fact that there is a bygone event that they really cannot get over; it takes command over their whole mentality. This people were the ancient Hebrews. They had been semi-nomads, moving a great deal in the desert, but having also certain periods in rather better areas where they could grow a bit of something. Like semi-nomads in general, they had longed to have land of their own, a settled land which they could properly cultivate. This is what they expected their God to provide for them, and what he promised to provide. Indeed the semi-nomads would tend to judge his effectiveness as a god by his ability to carry out his promise. The ancient Hebrews, the Children of Israel, had to wait a long time for their due reward, and perhaps this was the reason why they were so tremendously impressed when ultimately the Promise was actually fulfilled.

The earliest thing that we know from sheer historical evidence about these people is that as soon as they appear in the light of day they are already dominated by this historical memory. In some of the earliest books of the Bible there are embedded patches of text far earlier still, far earlier than the Old Testament itself, and repeatedly they are passages about this very thing. Fresh references go on perpetually being made to the same matter throughout the many centuries during which the Old Testament was being produced, indeed also in the Jewish literature that was written for a few centuries after that. We are more sure that the memory of this historical event was the predominating thing amongst them than we are of the reality, the actual historicity, of the event itself.

What they commemorated in this tremendous way, of course, was the fact that God had brought them up out of the land of Egypt and into the Promised Land. In reality it seems pretty clear that some of the tribes of Israel did not come into the land of Palestine from Egypt at all. Nevertheless I think it would be a central view amongst scholars that some of the ancient Hebrews came to the Promised Land from Egypt, and the impression of this was so powerful that it became the common memory of the whole group of tribes which settled in the land of Canaan; it became the accepted tradition even among the tribes that had never been in Egypt. Moreover the common tradition was the very thing that became the effective bond between the tribes of Israel, helping to weld them together as a people. This sense of a common history is always a powerful factor in fusing a group of tribes into a nation, just as Homer made the various bodies of Greeks feel that they had had a common experience in the past, a consciousness that they were all Hellenes. All this was so powerful with the Children of Israel because they felt such a fabulous gratitude for what had happened. I know of no other case in history where gratitude was carried so far, no other case where gratitude proved to be such a generative thing. Their God had stepped into history and kept his ancient Promise, bringing them to freedom and the Promised Land, and they simply could not get over it.

This was not the first time in history that gratitude had been a factor in religion, for at a date earlier still there are signs amongst the Hittites that the very sincerity of their feeling of indebtedness added an attractive kind of devotion to their worship of their pagan deities. But this gratitude was such a signal thing amongst the Israelitish people that it altered the whole development of religion in that quarter of the globe; it altered the character of religion in the area from which our Western civilisation sprang. It gave the Children of Israel a historical event that they could not get over, could not help remembering, and in the first place it made them historians—historians in a way that nobody had ever been before. The ancient Hebrews worshipped the God who brought them up out of the land of Egypt more than they worshipped God as the Creator of the World. By all the rules of the game, when once they had settled down in the land of Canaan and become an agricultural people, they ought to have turned to the gods of nature, the gods of fertility, and this is what some of their number wanted to do. But their historical memory was too strong. Even when they borrowed rites and ceremonies from neighbouring peoples—pieces of ritual based on the cycle of nature, the succession of the seasons—they turned these into celebrations of historical events, just as I suppose Christianity may have turned the rites of Spring into a celebration of the Resurrection. The Hebrews took over circumcision, which existed amongst their neighbours, but they turned even this into the celebration of a historical event. A Harvest Festival is an occasion on which even amongst Christians to-day we call attention to the cycle of the seasons and the bounty of nature. But amongst the Children of Israel at this ceremony you handed your thankoffering to the priest and then, if you please, you did not speak of the corn or the vine—you recited your national history, you narrated the story of the Exodus. It was set down in writing that if the younger generation started asking why they were expected to obey God's commandments they should be told that it was because God had brought their forefathers out of the house of bondage. Everything was based on their gratitude for what God had done for the nation. And it is remarkable to see to what a degree the other religious ideas of the Old Testament always remained historical in character—the Promise, the Covenant, the Judgment, the Messiah, the remnant of Israel, etc.

Yet this Promised Land to which God had brought them and on which they based a religion of extravagant gratitude was itself no great catch, and if they called it a land flowing with milk and honey, this was only because it looked rich when compared to the life that they had hitherto led. In the twentieth century Palestine has demanded a tremendous wrestling with nature, and if one looks back to the state of that region in Old Testament times one cannot help feeling that Providence endowed this people with one of the riskiest bits of territory that existed in that part of the globe. They were placed in an area which had already been encircled by vast empires, based on Egypt and on Mesopotamia and on a Hittite realm in Asia Minor. And, for all their gratitude, they were one of the most unlucky peoples of history. Other great empires soon arose again in the same regions, and they were so placed that they could not be expected to keep their freedom—their independence as a state only lasted for a few centuries, something like the period between Tudor England and the present day. The one stroke of luck that they did have was that for just a space at the crucial period those surrounding empires had come into decline, and this gave the Hebrews the chance of forming an independent state for a while. They virtually stood in the cockpit in that part of the world, just as Belgium stood in the cockpit in Western Europe and Poland in Eastern Europe. The fact that the Hebrews became, along with the Greeks, one of the main contributors to the formation of Western civilisation is a triumph of mind over matter, of the human spirit over misfortune and disaster. They almost built their religion on gratitude for their good fortune in having a country at all, a country that they could call their own.

Because of the great act of God which had brought them to Palestine they devoted themselves to the God of History rather than to the gods of nature. Here is their great originality, the thing that in a way enabled them to change the very nature of religion. Because they turned their intellect to the actions of God in history, they were drawn into an ethical view of God. They were continually wrestling with him about ethical questions, continually debating with him as to whether he was playing fair with them. Religion became intimately connected with morality because this was a God who was always in personal relations with human beings in the ordinary historical realm, and in any case you find that it was the worshippers of the gods of nature who ran to orgies and cruelties and immoralities. In fact, the ancient Hebrews developed their thought about God, about personality, and about ethics all together, all rolled into one. Because these things all involved what we call problems of personal relations they developed their thought about history step by step along with the rest. For a student of history, one of the interesting features of the Old Testament is that it gives us evidence of religious development from very early stages, from most primitive ideas about God, some of these ideas being quite shocking to the modern mind. Indeed, in some of the early books of the Bible there are still embedded certain ancient things that make it look as though, here as in no other parts of Western Asia, the God of History may at one stage have been really the God of War.

So far as I have been able to discover—approaching the matter as a modern historian, and rather an outsider, and using only what is available in Western languages—the Children of Israel, while still a comparatively primitive society, are the first people who showed a really significant interest in the past, the first to produce anything like a history of their nation, the first to lay out what we call a universal history, doing it with the help of some Babylonian legends but attempting to see the whole story of the human race. Because what we possess in the Old Testament is history as envisaged by the priests, or at least by the religious people, it is also a history very critical of the rulers—not like the mass of previous historical writing, a case of monarchs blowing their own trumpets. The history they wrote is a history of the people and not just of the kings, and it is very critical even of the people. So far as I know here is the only case of a nation producing a national history and making it an exposure of its national sins. In a technical sense this ancient Hebrew people became very remarkable as writers of history, some of their narratives (for example, the death of King David and the question of the succession to his throne) being quite wonderful according to modern standards of judgment. It was to be of momentous importance for the development of Western civilisation, that, growing up in

Europe (with Christianity presiding over its creative stages), it was influenced by the Old Testament, by this ancient Jewish passion for history. For century after century over periods of nearly 2000 years, the European could not even learn about his religion without studying the Bible, including the Old Testament —essentially a history-book, a book of very ancient history. Our civilisation, unlike many others, became historically-minded, therefore, one that was interested in the past, and we owe that in a great part to the Old Testament.

II

It was to be of great significance in European history that the Old Testament people did more than celebrate so fervently God's carrying out of his Promise to them. Their whole idea of history was based on the Promise; they believed in God's continuing Promise—a Promise repeatedly renewed but getting better every time. This is one of the earliest inklings that we possess of the idea which in modern times was to turn into what we call the idea of Progress. The Promise was always conditional; it depended on the Children of Israel behaving properly and remembering what God had done for them in the past. If they failed in their duty here, they would be visited by Judgment instead, a terrible national punishment for their ingratitude. Still, even after the most violent acts of judgment, God would renew the Promise to them, but carrying it each time to a higher level, holding out the hope of something better than ever in the further future. For these ancient Hebrews believed that God had not withdrawn his hand after creating the world. He could still do new things, they said; he could still do new things in history, things that were tantamount to a new act of Creation. When the ancient Greeks came along later they had a very different view, what we call the cyclic view of history—the notion that history is an eternal round of pointless repetitions, meaningless ups-and-downs, nothing ever happening that had not happened somewhere before. When you get to the book of *Ecclesiastes* in the Old Testament and you read that there is nothing new under the sun, that is not really an ancient Hebrew idea; it's the work of a writer who has been influenced by the Greeks. The Hebrews in fact did not hold the cyclic view. They held something like our modern view: that history is linear, irreversible, unrepeatable, and history is really going somewhere, the whole of Creation is moving to something that has never happened before. History is based on the Promise but the Promise may be punctuated by acts of terrible judgment.

Of course, what the world chiefly remembers about the Old Testament is this latter side of the picture—the hand of God operating as Judgment in history. No doubt the magnificence of the great prophets and the tremendous majesty of their literary work is partly responsible for that. Actually, if this had been the main idea of the ancient Hebrews, it would not have been an original one. The earliest interpretation of history that I ever read was produced in Babylon hundreds and hundreds of years earlier still, and even this only reproduced what had been believed in ancient Sumer at a date far earlier again: namely, that a national disaster or a defeat or a battle is the penalty for some neglect of the gods. They even discovered in ancient Mesopotamia that this theory will not work, it will not even cover the facts. They found they had to take the line that God sometimes passes over the actual culprit and visits the sin upon the children to the third and fourth generation. In ancient Mesopotamia you find men already wrestling with their gods about the fairness in the distribution of the punishment. Amongst the Hittites, a few hundred years before there was a Hebrew nation, there are most curious and moving debates with the gods, some of them anticipating in an interesting manner the wrestlings that the Children of Israel had with their God. It worried the Hittites, for example, that a King might commit a very serious crime but the gods might allow the culprit to live out a successful and happy life and then might visit the sin on the children to the third and fourth generation. And the people who received the penalty sometimes found it quite a business to discover what sin they were being punished for.

The question of a judgment of God that is being worked out in the course of historical events is not an easy one, therefore. I think all of us will be conscious of having done things, the memory of which makes us blush to the very roots of our hair because they would not stand being brought out into the daylight; they were more than an offence against one's better nature, rather like giving a stab at the very principle of goodness, a thing repellent to any observer and antithetical to the whole nature of God. On the other hand, if we ever actually try to picture God or pictorialise him—a thing we ought never to do (the Old Testament was wonderfully right about that from the very first)—it would be unworthy to picture him as sitting on a judgment seat. Some people would hold that sitting in judgment on men's sins is not a first-class occupation for a first-class God. The idea that history is based on the Promise, with a future opening out to something new, seems to me to be highly original, but in the teaching that national disaster is the result of a judgment of God the Old Testament picks up from earlier stages of human reflection and experience, from the pagans in Western Asia. It starts first where the Hittites left off. What is more interesting perhaps is that some of the more exciting parts of the Old Testament portray a God who has a distinctly gentle side in dealing even with terrible offenders. It contains many things that helped to civilise the world by encouraging even human beings to be a little merciful in handling one another's sins. And even there, the Old Testament isn't quite original. A Hittite King could curb his anger by saying, "After all, we are all sinners"; a Hittite King could entertain an idea of overcoming evil with good; a Hittite King in a prayer once reminded his God that even human beings

forgive the sins of a servant who has confessed and shown re-
pentance. The Old Testament is more definitely original, I think,
and, surely absolutely inspired, when it portrays God drawing
sinners to him, drawing sinners with the cords of love.

If in a sense the leading idea of the Old Testament is the notion
of history as based on God's Promise, a great section of this part
of the Scriptures does deal with judgment in history. And if in
the case of the Promise the thinking springs out of the first great
set of historical events (the exodus from Egypt and the entry
into the Promised Land), the notion of judgment is particularly
connected with the other great chapter in the nation's history—
Israel's defeats at the hands of surrounding empires, their break-up
as an organised people, the loss of their capital, the destruction of
the Temple and the exile of leading parts of the population especially to Babylon. I have already mentioned the fact that, looking at this story from the point of view of a modern historian, I
would have said that since vast empires had previously existed
in the neighbouring regions—in Egypt and Mesopotamia, for example—it was extremely likely that the same conditions would
produce the rise of further empires, and the Israelitish people
would not have much chance of keeping their independence.
From the point of view of the modern historian, I believe we
should say that the Children of Israel were very lucky (or favoured by Providence) to have had just a few centuries in which
to establish a monarchy and build up their own system. I do not
think that the modern historian would run naturally to the view
that God allowed or ordained the creation of a vast and powerful
Assyrian Empire for the specific purpose of just punishing the
Hebrew people for their sins. Of course one is still left with the
suspicion that if this Hebrew people had taken a little more notice of some of its prophets, they might have been saved from
some of their mistakes, and their disasters at any rate might not
have been quite so cruel.

But much though I shrink from the idea of judgment in history, I always have to come round to the view that it does exist. One of the most impressive things in favour of the idea is the fact that not only the ancient Hebrews but also the ancient Greeks (who differed from them so radically) insisted on something of the sort. One of the very big issues to which the historical mind has perpetually addressed itself in both ancient and modern times is the question of the decline of states and the downfall of Empires, including the curious question of the way in which liberty can be lost in a country that has once been free. Even when non-religious people have gone into this issue—even when a mind of cold steel like that of Machiavelli has gone into the matter—the conclusion is often a moral one: the excessive play of human egotisms, the lack of public spirit, the throwing off of moral restraint, can lead a society to disaster. The cupidities of the human being can get so out of hand that they are too much for the managerial power of any statesman. The Old Testament idea of judgment is one that has reference not to the problem of individual suffering but to the fate of great collectivities— the fate of nations, cities, organisations, even universities, even churches. For many centuries of the past, and down to pretty recent times, men were much better aware than we have been of the fact that society and civilisation are built on a volcano. If there ever comes a point at which too many people become too wilful the situation becomes too hot for a statesman to handle by the sort of political adjustments that are in his power. When Ayub Khan retired a few years ago he announced something of this sort. He said, "Every problem is now being solved in the streets. Mobs surround any place they like and force acceptance of what they like." In quiet and settled times people always forget how easily a society can come to breakdown and a civilised world collapse. If a society even becomes too materialistic in its objectives, so that the competition between its members for more than their due share of the materialistic benefits gets too intense, then I think there is a judgment of God which is embodied in the very constitution of things which will bring that society to total collapse. That is how the world is constituted. In this sense there is a Providence, there is a judgment that is embodied in the very constitution of the universe.

I have always said that it was a great pity that after the war of 1914 the Germans did not ask themselves, "What did we do wrong?," instead of complaining so bitterly about the unfairness of their fate, and landing themselves, partly through too great resentment, in the hands of Hitler. But it was necessary for Englishmen to do the same about the war of 1914, necessary for them to ask, too, what they had done wrong, not imagining that because they were on the winning side in that war God had adjudged them to be in the right. And certainly it was not for the Englishman to pretend to be an impartial judge and to assess the moral responsibility for a war in which he himself had been one of the belligerents. Something is achieved if men can be induced to pass judgment on themselves and to see where they have been wrong, induced to see how their actions would appear to God. In fact I think we would really have to say that the war of 1914—the source of so many later horrors and barbarities—was itself a judgment on Europe, but a judgment on our whole civilisation at that time. Even so, the judgment of God is never so monstrous as the judgment of men when they start behaving like gods and judging one another. The Old Testament at its best is full of the idea of mercy.

In a certain sense it is true to say, I think, that both parts of the Bible—both the Old Testament and the New—are treatises on human suffering. They show every sign of being the result of long attempts to deal with the problem of suffering. The inhabitants of ancient Palestine knew suffering more than anybody else; they were just the people to be tormented with this problem, as they were beset with such colossal disasters. They wanted to know the meaning of it all. The Old Testament is packed with evidence that this is the thing that made them enquire so persistently about the meaning of history. And, in general, in those ancient times, people seem to have been more anxious about their destiny, more anxious to find the meaning of history, than the men of the twentieth century. I want to emphasise the fact, therefore, that the Old Testament does not rest satisfied with the view that suffering comes as a result of a terrible divine judgment. Indeed it is here that the Old Testament becomes most powerful and original. It reaches one of its highest spots at the point where it works out that the Jewish nation is suffering in reality for the sake of the rest of the world—its suffering is an essential part of the mission that it has to fulfil. They discover that if they were God's chosen people it was not in the sense of enjoying favouritism, not in order that they should have a specially good time. They were chosen to carry out this mission, chosen to bear the special suffering that it involved. Through their suffering they carried religious thought to greater depths than ever before and turned what had originally been to them their local God into a God solicitous for all mankind. Through their sufferings they would help to draw the rest of the world to God, partly because even exile and depression introduced them to a wider world. At some of the loftiest moments of the Old Testament they reached the idea of vicarious suffering—Israel voluntarily accepting the suffering, they themselves taking it on instead of the other people of the world, reconciled to it because they saw it was leading to something, saw that it had a purpose to it. So far as I know this is the first example on record of a nation having a historic mission, this being, in the case of Israel, the mission to teach the world the religion of their God, Jehovah. It is the only example I know of where a nation had a historic mission that did not in itself involve—or didn't *need* to involve—aggression and aggrandisement. A modern historian would have to say that they were right in this piece of interpretation—this was the historic role of the ancient Hebrews. Later, their role was to spread the teaching of their own scriptures over the Roman Empire, as far as Great

Britain, and ultimately as far as San Francisco at the opposite end
of the world.

III

One of the most unexpected aspects of the story is the fact that
the ancient Hebrews—having been more obsessed with history
than any other nation that I have ever come across, more even
than England with its doting on *Magna Carta*—made a curious
change in mid-stream and, in a way, ceased to give history the
priority. It is one of the things that happened after the crowning
catastrophes: the break-up of the Hebrews as an organised people, the loss of Jerusalem and the destruction of the Temple, and
the exile of so much of the leading population to Babylon. The
faithful remnant who were able later to return to their own
country and pick up the threads of their national life were convinced that their disaster had been a judgment from God, a punishment for disobedience. I wonder sometimes whether now, indeed, they had not learnt the lessons of history almost too well.
They determined that, henceforward, they *would* put the commandments first. It is at this point that there begins to emerge a
Judaism distinctly different from what had previously existed,
because henceforward the obsession was for the Law rather than
for History. In a sense the people now began to use the Old Testament itself not for the sake of its record of God's historic acts,
his doings and transactions with human beings, but rather for the
sake of the Law which could be extracted from it, the Law in
the strictest literalistic sense. Indeed, they seem to have partly
re-edited, partly re-written, a good deal of the scriptural history
for the purpose of emphasising this legalistic point of view.

It had been held that part of the sins for which they had suffered so terrible a judgment had consisted in their consorting
with foreigners, and becoming contaminated by the idolatries of
these strangers. Now they went into extreme reverse, closed the

ranks against outsiders, forbade mixing with the foreigners, paid great regard to racial purity and, in a way, set out to insulate themselves against the rest of the world. Far from saving them from a repetition of disaster, the new policy helped to bring still further catastrophes upon them. But, this time, they had been so careful about the commandments; they knew that they were not open to the same reproaches as in the earlier history in the days of the great prophets. It is interesting to see that even pious people were not now so ready to interpret their disasters as a judgment for sin. Sometimes, in most moving passages, they reproached God for persecuting people who tried to be so faithful to him. He seemed to be letting the pagans and the idolaters win every time, they said. One even finds them suggesting to God that he is not doing his own cause any good by this.

I think that the historian, looking at the narrative from the present day, and making his verdicts without any religious prompting, would say that in that sector of the globe empires were coming to be still more colossal, still more formidable—the Persians, the Empire of Alexander the Great, and then the Roman Empire. We today can hardly see how the Jews of this period could have expected to be able to keep their political independence, owing to their very situation on the map. At the same time one might wonder on occasion whether the Jews always showed sufficient political wisdom, whether their sufferings might have been less severe if they had been more worldly-wise. One of the difficulties in the Old Testament period is that the question of political wisdom is made to appear so irrelevant and is so difficult for us to pass judgment on. One can possibly say, however, that, though their fanaticism may have led to some political mistakes and to mundane disasters, it helped to achieve the ultimate preservation of their religion, and it helped them to maintain their identity as a separate people down to the present day. In one respect the predictions of the prophets proved to be fairly correct—so many of the peoples in that part of the globe simply disappeared from the map, their destruction as organised peoples proved permanent. And the persistence of the Jewish people down to the present day is one of the amazing aspects of the story.

At this late stage in their history—after what we normally regard as Old Testament times, and in the very latest centuries before Christ—the Jews produced another signal contribution to the study of man and his destiny. In the changed circumstances the notion of history as based on the Promise received a new development. It developed implications which in a way made it inimical to the cause of history; instead of being a help it became a hindrance. The great prophets of the period of the Exile had pointed to a fine future that was in store for the remnant of Israel, the faithful few who would return home after their life in Babylon. But, as we have seen, that reward never materialised. The Jews remained the prey of great empires in their vicinity, and the mundane success, at any rate, did not come to realisation. Indeed, as the Roman Empire grew in power the general structure of the world made it pretty clear that the Jewish people could not expect a genuine improvement in their position in the ordinary course of things. Despairing of any future in the ordinary historical realm, they began to look beyond what we normally regard as history. They looked for the end of all the ages when a dramatic act of God would rectify their position once and for all. They turned from history to dreams of a messianic kingdom which should exist in the world but would emerge as part of a supernatural order. They concluded that the end of history, the end of the world (in the ordinary sense of the words) was at hand and they spread themselves a good deal on what we call eschatological speculation. All this led to a kind of utopianism and even to some unfortunate adventures in the political realm. Instead of looking at the past they set their eyes on a future which represented a great cataclysm, regarded as very imminent and occurring outside the ordinary realm of historical happening, outside the ordinary workings of politics. Instead of turning their minds to the God of history they looked to a God whose great accomplishment was to be something in the future.

There is something deeply inimical to history in this kind of day-dreaming with political unrealities. It led to the production of a considerable literature which appears between the Old Testament and the New and is sometimes disturbingly fanciful, though it is sometimes deeply moving. In a sense it is remarkable that it should have appeared amongst the Jews because in general we should say of their religion that it was very down-to-earth, very close to realities. In those ancient days it was a progress towards what we should call rationalism, because having a single God—and he a very jealous one—meant that you swept away the great mass of elaborate mythologies that existed in the pagan realm. This new literature, however—only a part of which appears in our Apocrypha—shows a considerable increase in what one might call the mythological element. It is also paradoxical because, at a time when the Jews had closed in on themselves and hardened in their exclusiveness and set up barriers against the outside world, they did in fact begin to be influenced in their ideas more than before by the outside world, for example, Persia. In this period of apocalyptic writing they do seem to have influenced historical study in one interesting respect. Dealing with the Last Things—the end of ordinary human history—and thinking of the successive stages that must lead to the End, they started the practice of periodising history, dividing the continuous stream of history into stages or periods that could be named. Indeed their way of actually doing this still had its influence in Europe down to the beginning of the eighteenth century when the division into ancient, medieval, and modern began really to prevail. Also, in spite of all that I have said, the Jews did not lose the art of writing history. From the point of view of a modern technician the First Book of Maccabees, written not long before Christ, is one of the greatest historical narratives ever produced

in antiquity. Furthermore, in regard to all the messianic and apocalyptic literature that I have been describing, we have to say that,
of course, Christianity was affected by it, as well as by other,
more deeply spiritual, movements that were taking place in Judaism.

But in the literature that is in question here, the Jews were the
forerunners of something else that keeps reappearing in history—
the phenomenon of political messianism. We now know that it
can exist in a secularised form, apart from religion, and it has
existed often enough for us to be able to examine it almost scientifically. It seems to be the product of a feeling of desperation
amongst well-meaning people who have come to regard the
world or society as too evil to be coped with by ordinary political
methods or by the operation of reason. There must be a cataclysm, therefore, and a blood-letting, a final show-down between
good and evil. There was a wing of the Reformation, to which
we owe some very profound and beautiful ideas, and yet the
very people capable of these could say "Kill, kill, kill"; they
really felt that the only thing to do was to annihilate the wicked.
That is one of the reasons why the most formidable result of the
Reformation was a colossal increase in the power of the state.
The world ultimately becomes weary of the disorders. Men want
the return of a normal reasonable life so that they can pursue all
their non-political objects. They throw themselves into the hands
of a Messiah, a saviour of society; and in the sixteenth century
the despotic monarch is greeted as the Messiah. Once there is an
appeal to force it is the wrong party that ultimately brings up the
bigger guns. The wicked can go further in unscrupulous violence
than the righteous will care to do when it comes to the point.
The cause of goodness depends on the slow growth of reasonableness among men, and it was one of the virtues of the early
Church that its leaders saw how the cause of Christianity itself
needed peace and settled order in which reasonableness can operate. The concept of a Messiah or a Messiah-King, unless it came

to be interpreted only in a higher spiritual sense, was bound to be one of the most dangerous and self-destructive concepts ever produced. I should like somebody to convert me on this one great point. But, so far as I can see, the Old Testament itself is weak on the political side, and is of such a nature that it deters one from thinking about things politically; even its commentators seem to give us too little help on this side. I would like to know how wise the prophets were as political advisers. I suspect they were right, for example, when they cried out against reliance on Egypt as an ally.

But I have not exhausted the contribution of the ancient Jews to our world, or their originality in the realm of thought. Conceiving of God as a person who was very close to them, very closely involved in human affairs, they walked with him, they conversed with him, they wrestled with him on ethical issues, sometimes with a frankness that is quite startling. Above all, they established an intimacy with him, realised him as a spirit, knew that he was inside them, and that if you plunged to the depths of the ocean you would find him there as well. It is this God who exists in the very inwardness of things that we have to have in mind when we think of God in history. Even in the prophets, it is not the theory of judgment that is most impressive of all, but the whole dialogue between God and man that is associated with it, the whole picture, for example, of an entreating God who tugs at the hearts of men. It was partly out of the magnitude of their sufferings that the ancient Hebrews reached such heights and depths in their relations with God. And even after the days of the Old Testament were over, the extraordinary spiritual intensity continues, bringing us to the brink of Christianity. No historian can ever doubt the colossal historic effects of these spiritual experiences. In the long run the whole issue of religion depends on one's judgment of them as authentically spiritual—it depends on the validity, the convincing character, of man's inner experience of God.

6

The Modern Historian
and New Testament History

I

It must be a matter of regret that the followers of Jesus did not provide for us something more like an ordinary biography of their leader, in addition to all the other things that have come down to us. I regard the myths associated with Bethlehem as the richest and profoundest that exist anywhere, but I wonder whether they did not prevent the early Christians from collecting the more concrete and factual things that concerned the childhood and early youth of Jesus. I sometimes tell myself that those people who have no use for the past still seem, as a rule, to be interested in stories about human beings whom they have known. As a rule they seem to be willing to go back at least to the tales of a grandfather. For this reason, I find it difficult to imagine the disciples failing to talk to one another and to other people about Jesus as an ordinary human being. And, supposing one were to hear for the first time about somebody who had risen from the dead, I find it hard to believe that one would not want to know many incidental things, and would not engage in a good deal of cross-questioning in order to learn about the mundane side of his life.

I can see why, in an ecclesiastical tradition, a great deal of this would ultimately disappear from the formal records, but I deplore this result and think that, in the kind of world that is developing today, this lack may even make the Gospels less plausible to anybody who is not already accustomed to them. I notice that in reality some people *do* feel that to some degree Jesus has been turned into a sort of lay figure—one of those "characters," those "human types," which essayists used to write about when they described the model of "the avaricious man," or "the rake" or "the unjust steward." Oddly enough, when I think of the Gospels in their historical aspect, I am often reminded of the stories which are told about Winston Churchill and the hundreds of sayings of his which have come into currency during the last twenty or thirty years. But along with the vast miscellany of anecdotes, we do have a consecutive and structured biography of Churchill, whereas (save perhaps for one very short period) there are hardly the materials for such a thing in the case of Christ.

A further question arises, however. That is the problem of the historicity of the stories, anecdotes, and sayings which have come down to us, appearing at first as an oral tradition, whether in the case of Jesus or Winston Churchill. So far as Churchill is concerned we may read about these things in narratives or books, but we know that they largely go back to an oral tradition and often must depend on the evidence of a single man. Sometimes indeed it requires only a little detective work to discover that a certain story must have been told by Churchill himself. He gloried in the neat repartee which would throw the other man off his balance, and he liked to tell the story of this afterwards. I would not care to be the guarantor of any particular one of these Churchillian anecdotes that have come into circulation. I am too aware of the fact that even modern historians whom I have known—men utterly reliable in their own technical field of study —seem to find it hard to retell a story at dinnertime without adding something to round it off or to give it a special piquancy. I suspect that it would not be easy to establish in a watertight manner the truth of a great many of the single stories about Winston Churchill. Even when a number of writers or a number of raconteurs produced the same anecdote, there is still likely to be considerable doubt. One must be prepared for the possibility that all of these went back to a single original faulty source.

I do not know that we need to be greatly dismayed by this fact, however. I have a hunch that many of these stories will have some truth in them, though it might be impossible for me to distinguish between the authentic and the inauthentic. In fact, the whole corpus of anecdotes might still reveal Churchill the man in a very effective manner. A Churchillian story that ultimately proved to be apocryphal might even be more true, in a sense, more typical of the real man, than some other story unquestionably authentic. The person who invents such a story, rounding it off in order to give it the true Churchillian ring, is just the man who will produce the sort of thing that contemporaries will recognise as typical. At the same time, in the course of actual life in the battles and the flurries of real existence, a man will often in reality produce an action or a saying that is not typical, but out of character. It might even be true that the anecdotes in the mass —whether all of them are authentic or not—can bring us closer to the man who is the subject of them than is possible with an historical work limited to things that can be proved by the right combination of documentary evidence.

I think it would be relevant if I were to insert a note here about the great trouble which the saints have given me in my studies of historical technique. They may be as truthful as they know how to be, but the point is that *that* is not enough. I think this is a thing which must touch our present problem. Over the course of centuries I think you discover that mere honesty is not enough; a mere desire to be honest is certainly not enough. If a man is to give truthful evidence, he must be capable of great internal discipline, capable also of a great clarity, all of a sort that is rarely achieved except by technical training. Throughout my adult career I have insisted that anybody reading secular history ought to be sure to have a Scotland Yard detective at his side. And I have said with equal consistency that those who read ecclesiastical history must have two of these detectives, because, in this latter case, a man so easily runs his theology into his evidence. In this I had chiefly in mind the writers on later church history, those, whether Protestant or Catholic who take a party line. I suppose Renan was the man who first raised the issue whether a believing Christian was the right kind of person to produce a good biography of Jesus. Some people seem to think that the ideal biographer would be a man who had once been Christian but had then abandoned the faith; presumably he could be impartial, able to see both sides of the story. I cannot agree with a policy that makes a lapse from the faith the essential qualification for writing religious history. However, I think it might be a good thing to have two biographies of Jesus, one of them written by a Christian, the other by a non-believer. Holy people, much as we may love them, can be quite exasperating for the historian, not (I think) because they fabricate untruths, but perhaps rather because they do not know how to distrust other people's reporting, and do not go far enough in analysing and questioning even their own impressions of things. Sometimes they think it is a virtue just to *believe* a thing that is enough to knock the bottom out of any serious historical study. I have known some saints who were over-trustful even in regard to stories or rumours of mundane things that had nothing to do with religion. The saints can sometimes be rather naughty about these historical matters. In the Middle Ages there are one or two pious writers who take the line that if they tell a story about one saint when that story really belongs to another, this is nothing for anybody to complain about, since in the communion of saints all are one.

II

About the crucial Christian narrative I think there are a number of things which the ordinary general historian would be likely to agree to. Because Christianity presided over the rise of Western civilisation, even the secular student cannot deny the tremendous importance of what happened in the Holy Land in the years —and indeed in the weeks, no doubt in the few days—immediately after the Crucifixion. That event seems to have found the disciples in great disarray—a frightened Peter wildly untruthful, for example, and denying his connection with Jesus. The description of their inadequacies must have come from the disciples themselves, for neither the authors of the Gospels nor anybody before them is likely to have had any motive for inventing such stories if they had not been true. The Christian religion, based on the belief in the Risen Christ, must have emerged at this time, and it seems to have acquired its familiar general shape with a rapidity calculated to fill the historian with surprise. I do not know whether I am wrong in thinking that any claim about Christ's rising after three days would have lost much of its force if the announcement had been long delayed. Supposing there were any serious doubt on this point, it would make a number of other things more uncertain.

Henceforward, the disciples embarked on the greatest missionary enterprise of all time. Though they had doctrines to preach, everything hung on the events they had witnessed, everything depended on the actual evidence they had to give, and on their words carrying conviction at this point. Indeed the whole type of authority later developed by the Church—the stress on tradition, for example, and the desire to keep to what was believed in the first generation—rested on this realisation that everything depended on the eye-witness evidence of the disciples. Jesus had left no written works, and though Eusebius later imagined that he had tracked down one of his letters, this was a great mistake. The puzzling thing to me is that not only for the lifetime of Jesus but also for nearly a generation after that, we lack what I might call the original written records, providing the evidence in its raw and uncooked state the way historians need to have it— that is to say, before it has been mixed with doctrines of a later date.

I personally feel that, in spite of this, we can know, roughly speaking, the strategic things that the disciples had to report. In this sense, whether the historical handling of these matters was bad or good, we today are in something like the position of most of the people living at the time—we can know the things to which the disciples testified, though we only hear of them at second hand. I am aware that some people do seem to doubt whether we really know what the disciples witnessed and what they preached at the very beginning. I suppose they believe that the original teaching has come down to us buried in a lot of theology which the Church may have developed later. That is certainly the kind of point that the critical historian has to bear in mind. In any case, we always have to remember that if a handful of us were to see with our own eyes and actually touch, even converse and actually exchange memories with, a man whom we could absolutely attest to have died and been buried the week before, the task of communicating our certainty to other people who had not been witnesses would be full of difficulty. Supposing the early Christians had indeed taken much more care about fixing the evidence in writing at an early date, this would not have eliminated or even lessened the obstructions to belief that exist today.

One gathers that in the early days of the Church many people believed in the preaching because of certain sanctions, certain external guarantees, that accompanied it. The disciples were convinced that they had seen the Risen Christ, and for them the Resurrection itself of course was a sufficient guarantee. But the Resurrection could operate in this way only for those who already believed in it, possibly only for those who had seen it for themselves. One gathers that a number of people accepted the preaching because of the miracles that accompanied it. Once again, however, unless one has actually seen the miracles for one-self they require as much of a guarantee as they confer; they even add to the difficulty of belief.

For a present-day historian there still exists what might be re-garded as further sanctions, certain external things which give some additional weight—I do not think one can say more—to the evidence that has come down to us. Firstly, there was the extraordinary transformation which seems to have taken place so quickly in those disciples who had been apparently so badly broken and demoralised at the time of the Crucifixion. Secondly, there was the fact of the extraordinary spiritual power, the profound internal depth, which a handful of humble men acquired, as well as the body of tremendous doctrine that so quickly emerged. Thirdly, there was the amazing missionary movement which braved the power of governments, occasionally prevailed against religions long entrenched, and carried men on in spite of the constant threat of imminent martyrdom. And at any rate it seems totally impossible to discover any personal ambitions or vested interests that could have induced the disciples and the preachers to carry out a sort of hoax; they must have known they were going to almost certain death.

In connection with the general question, What made so many people accept the Gospel?, one might add that many of those who heard the first missionaries may not have put the microscope on a single point by itself, such as the problem of the Resurrection. They may have accepted the whole of the preaching as a package deal, partly because perhaps it brought them to a higher conception of God, partly because it may have corresponded so well with their conception of the whole human predicament, and partly because they might see a lot of point in the call to re-pentance in any case. Adopting Christianity as a whole new composite view of things, they might even find the Resurrection less of a problem in itself, less of a paradox, once the message as a whole had been presented to them. Perhaps the momentum of the whole compound movement would carry them over the hurdle. I imagine that a number of people accepted the Gospel in the feeling that it offered a kind of deliverance from an imminent doom, the end of the world being at hand. Furthermore, once they had thrown themselves into the new religion, they gained the conviction that it had ratified itself in their own inner life; they became sure that they too had actually found contact with the Risen Christ. I personally attach very great importance to this point.

The sayings of Jesus, and the stories about him, were collected and circulated very soon, the true ones no doubt mingling with the false ones. The technical historian of the present day must regret that we cannot be sure that we possess them in the form in which they first appeared, with an indication of the actual source of each of them. All these materials, these reports of actions and sayings, went through a considerable amount of screening in the Church itself as time went on, of course, screening which in some ways assists the historian, though in some ways it makes the problem even more complicated than before. Interesting techniques have been developed by modern experts in biblical studies to elucidate what happened and in particular to discover what faults emerge in material that is handed down at first by oral tradition, what additions and distortions take place with the lapse of time.

This work by the divinity scholars the modern historian must greatly admire, I think, but there might be one point on which the secular student would put additional stress. All would agree, I imagine, that some of the things in the Gospels—the Crucifixion and the Resurrection for example—must have been in the preaching from the very start. This was the reason for everything else that happened, and the evidence must have come from the disciples. Indeed, I think there can be no doubt that the belief of the disciples in these events was of an overpowering nature. Then, again, I think one must agree that, from the state of our available sources, it looks as though the developing tradition paid particular attention to the Passion of our Lord, the stages on the way to the Crucifixion. Here, as one can see from the Gospels, there developed (and apparently at an early stage) the most detailed, most consecutive, and most consistent piece of narrative about Christ that got into the transmission process and was properly handed down to us. It has been pointed out that one or two incidents in this chapter of the story are included for no discoverable religious purpose, no propaganda reason, except possibly to call attention to certain outside people who had been identified as eyewitnesses. It is also rather impressive to see how full and precise (compared with most things in those days) the story of the Crucifixion is, and how reticent in general the Gospels are about the manner of the Resurrection, a topic on which the imagination might have run riot. The sayings of Jesus were obviously an important matter from the very start, and I imagine that collections of them might have been making their appearance pretty early. In spite of all this, my one warning would be that some of the anecdotes and sayings most likely to have come from the disciples themselves must have kept on appearing in sermons or must have been taken into the current practice of the Church. These would be the ones most likely to survive in the long run, but they might become a little bent with the passage of time, especially when used for sermons or catechetical purposes. There is no more dangerous cause of distortion than religious preoccupations or the telling of stories for the purpose of the lessons that can be drawn out of them.

I suppose that our four Gospels—which appear, in their present form at least, rather late—gathered together all the materials then available, all that had gone through the Church's screening process—the thing which no doubt explains why they became "ca-

nonical." From what is known of a considerable mass of apocryphal stories and sayings of Jesus, the Church at various periods
showed remarkable discrimination in what it accepted and what
it rejected.

I personally have tended to feel that the recognised tradition
concerning Jesus must have reached its familiar shape fairly early
—possibly within something like twenty years of the Crucifixion
—while many witnesses were still alive and there was every
chance that false evidence would be challenged. But the Gospels,
appearing fairly late, do not confine themselves to telling us
about the things actually remembered by the disciples. One can
see from the reading of them that they are born of something
more than the straight recollection of mere facts. The course of
events is being reinterpreted in the light of things that happened
later. The significance of Christ is being affected by developments
that took place after his death. In any case, the Gospels are not
biographies in the modern sense of the word; they are not close
studies of the environment, the education, the early influences,
the intellectual development, the all-round personality of the
man. This kind of thing could not have been expected for anybody in those days, though something closer to it would have
been definitely possible. What we possess in the New Testament
is the story of Jesus told with the object of communicating at the
same time a particular view of him. And, in providing this, the
Gospels do not differ from a lot of historical writing, a lot of
the literary sources that even the modern historian has to handle.
Possibly it is a more serious matter to have to recognise the fact
that the Gospels are governed by a strongly dynamic religious
purpose. I think that in any kind of literary or historical work
such a religious purpose tends to operate with a powerful transforming effect on the materials at its disposal. Technical historians are bound to see a danger here. That is why one must
wish that we had had the evidence of the disciples—the memories
of the disciples—in the original relatively uncooked state, though,

of course, we must never forget the limitations of all human evidence, even that of eyewitnesses in an ordinary motor accident; the disciples—whether consciously or unconsciously—may have given a subtle twist in any case to the evidence they had to offer.

III

There is one thing that still worries me. I always think that the early Christians may have found the most interesting technique ever devised for evading the essential historical issue and for securing that they should not become historically minded. They came to identify Christ with the concept of the Messiah and with the picture of the Suffering Servant in the fifty-third chapter of Isaiah. They found a number of other things in the Old Testament obviously applicable to him, and enriching their view of him. All this threw light on things which Jesus had said about himself and which they may not have properly understood at the time. For them everything now fitted into place, and they began actually hunting for parallelisms and analogies in the Old Testament. Sometimes indeed the quotations they found were too farfetched, not likely to convince the historians of the present day. Sometimes it meant giving passages a meaning that would not have been recognisable to their original authors or to the people living in Old Testament times. They found texts which the Jews themselves had never regarded as having anything to do with the Messiah. Nobody on earth could have understood some of these passages as prophetic until some date after the prophecies alleged to have been embodied in them had actually been fulfilled. It was pointed out that some of the supposed prophecies which they found were not in the accepted text of the Old Testament at all— a point which, I gather, gave Churchmen a certain amount of worry at a later time. The Old Testament in its aspect as a mere record of events that had happened even came to be depreciated— divine inspiration not being needed for anything so pedestrian as mere narrative. Justin Martyr, in the middle of the second century, considers the Old Testament to be inspired because it is prophecy, but he leaves the suggestion that the books which we regard as the New Testament did not need to be inspired—they were only accounts of the fulfilment of the prophecies. Indeed, in a way, the whole of the Old Testament came to be transmuted into prophecy, regarded as a prefiguring of Christ. The Hebrew Scriptures were taken over by the Church but used now to an entirely different purpose. Their historical aspect was not denied but was depreciated. In any case this entire project entailed discovering figures of speech and even concealed prophecies in everything, turning the Flood and the passage of the Red Sea into tremendous allegories, for example. Furthermore, so far as the earliest Christians were concerned, one of the most important aspects of the Gospel was not so much the question of what Christ had done during his life on earth, but rather the demonstration that he had fulfilled all that the prophets had said about the future Messiah.

All this is found today to raise the question whether this whole programme could not be made to operate in reverse—Christians coming to believe that because a thing had been predicted it must have happened, indeed must have happened just in this period of Christ's life on earth when all the prophecies were supposed to be coming true at the same time. In any case the attempt at the crucial period to recover (or reconstruct) a portrait of the personality of Jesus might be misdirected, too much dominated by certain brilliant passages from the Old Testament which perhaps would transform the very memory people had had of him, altering some of the lines of the picture that was finally produced. In respect of one or two points even in the Crucifixion narrative one may have to ask whether some little pieces of history might not have been inferred from the prophecies instead of being guaranteed by confirmable evidence.

It is interesting to see how the prophecy could sometimes be treated as though it were a sufficient proof of the actuality of a given event—the thing must have happened as a real fact, for was it not foretold in the Old Testament? And, though it may have been merely a linguistic usage (so that I would not propose to stress the point overmuch) do we not read that such-and-such a thing happened in order that the prophecy, the Scriptures, should be fulfilled? If you really believe that there was a mechanical connection, an absolutely reliable connection, between the prophecy and the fulfilment, and if you believe that in Jesus all the prophecies of the Old Testament were being fulfilled virtually at once, the use of prophecy as a form of historical evidence can be absolutely honest and sincere. Indeed, precisely because of its divine character it was a better source for history than historical evidence as we understand it. In the work of Justin Martyr, Jesus comes to be described as "deformed" in order that he shall correspond with the second verse of Isaiah 53. Justin also takes the line that, whereas one's reporting of an event might be questioned, the establishment of the fact that had been prophesied constituted testimony of a more absolute kind. I wonder whether, for the ordinary secular historian of today, this might not be one of the most delicate issues that early Christian literature presents.

At a later date, even St. Augustine makes a curious and disturbing remark. He says that the Old Testament must be accurate in its history because its prophecies have so often come true. One finds the recurrent assumption in the early Christian centuries that inspiration would not have been given to the writers of the ancient Scriptures merely for the pedestrian task of recapitulating the things that had happened in the past. The real voice of God was in the prophecy of Christ. Historical narratives would be turned into allegories, the Flood and the passage of the Red Sea into prefigurations of something else. In all this the early Christians sometimes affronted their respect for the sheer concrete facts of history, that respect for actuality and solid truth, which the modern historian regards as so important. Only gradually during the first three centuries did Christians—so intent rather on the spiritual life—fully realise the historical aspects of their faith, the sense in which our religion is peculiarly tied to history.

I am not sure that, in the earliest centuries, there was not a danger of Christianity being unhooked from ordinary history, too disconnected from hard facts, much as this contradicted what later (and for ever afterwards) became the essential attitude of Christians on this point. Sometimes, when they debated with the Jews or when they studied the ancient Scriptures, they would fall into an attitude which the modern scholars would describe as "unhistorical," an attitude not unlike that of those Marxists who carry so much of their present into their reading of the past. Here, in point of fact, is something which is liable to happen to religious people, and a parallel deprecation of mere history can seem a natural thing to those who have decided to put their heart and their treasure in Heaven. One can quite understand that, for those to whom the saving of souls is an urgent matter, there might not be a great deal of point in sitting and brooding over bygone times. Such a thing would be still more likely to happen if the early Christians believed that the end of the world was at hand—a prophecy singularly inaccurate in the sense it was construed at that time (though there might be something to be said for our always living as though the end might be at hand). And, as a result, the early Christians might feel that after Christ's victory no further history could be really meaningful: ordinary human events could not matter any more.

Let me make one thing clear, however: history may be a matter of importance to human beings, but we must never make it all-important. And the question of our religion must never be regarded as merely a matter of historical evidence. Sometimes the two do come together, of course. I am never quite happy about the view that the people of Israel were brought to exile and enslavement merely as a result of their sins, and I notice that some rather touching writers of theirs in a later period themselves reacted against this view. When an alternative interpretation was offered—when it was suggested that God did not really choose Israel for prosperity, but actually chose this people for suffering, because through this all the world would be brought to hear of the Divine Father—this seems to me not only good religion but also a first-class historical statement which, in a fundamental sense, a twentieth-century historian can accept.

Finally, let me stress a point which should now be obvious to all readers of this essay, and which is of first-class importance. Those who write history, or talk about it, must include in their narrations many things which have not been proved in an absolutely watertight manner. People are very loose in their reporting sometimes. It is wrong to assume—as people so often do when they are dealing with the Scriptures—that if an event has not been demonstrated with mathematical certainty, it has been proved not to have happened at all.

7

The Establishment of a
Christian Interpretation
of World History

I

Christianity is linked with history because it hinges on one or two events which, though possessing what might be called a transcendental character, are also fastened down to a certain place and a certain date. On these events—particularly the Crucifixion and the Resurrection—the disciples seem to have concentrated their chief attention. It would appear to be the case that the disciples acquired authority in the Church not exactly as theorists in theology, but as men who had been eyewitnesses. St. Paul himself seems to have felt that he had some authority because, through the vision that he had had, he too could claim to have been in a certain sense an eyewitness. On the high matters that are here in question we today are in something like the position of those who heard the first Christian preaching, for unless you are an eye-witness yourself you are bound to be dependent on somebody else's evidence. If the first disciples and early Christians had been trained historians they could hardly have altered the predicament in which we stand at the present day. For a lot of the more mundane things, which would interest the secular historian—what Jesus looked like, who were his teachers and boyhood friends, and how are we to date the successive stages of his mundane life—we are liable to be defeated, because the early Christians were too other-worldly, and all the stories about him were sifted and sieved through the machinery of an ecclesiastical tradition. Many of these stories would be lost because they served no particular religious purpose; some of them would be lost no doubt because it is often so difficult to trace such things to an authoritative source. Indeed when we are thinking of secular history in general we do perhaps have to ask ourselves the question why spiritually-minded men—men who put their heart and their treasure in heaven—should worry about so mundane a thing. The question is very relevant. The Christians did come to be greatly interested in the past; indeed it was to be Christianity that helped to make the West so historically minded. Perhaps the key to the whole story is the fact that Christianity was not a religion centering on some vague demiurge or some figure taken from Egyptian or Greek mythology. It had one foot on the hard earth; it had a definite linkage with history. In the theological conflicts that occurred in the Roman Empire there would be parties—the school of Antioch, as against the school of Alexandria, for example—which insisted on not losing sight of the humanity of Jesus, not letting it be submerged in his divinity. In view of this, it was natural that men should want to know about his human life. Also you were bound to become interested in the fact that He came at a certain date, at a certain point in the time-series. Granted that all this was to happen, why did it happen just then? Pagans asked, Why had God waited so long before revealing Himself in Christ?

In any case, once it was granted that at a crucial moment Christianity was vitally concerned with history, the implications of the commitment were bound to expand as a result of the questions coming from the outside world. Once the Church had become a sedentary affair, an enduring institution, conducting missionary work and propaganda in a pagan world, it became necessary to answer the charge that Christianity was a recent innovation, a religion only brought into existence, so to speak, the other day. This meant insisting on the continuity with the Old Testament while stressing more than ever the divergence from Judaism. The case for the Church was enormously enhanced by putting forward a historical claim of a very imposing kind: Christianity, the true religion of Yahweh, was older than Judaism itself, older than the religion as established by Moses. In fact Christianity regarded itself as the recovery of the original religion as it had existed in the more primitive ages of the world—in the time of Noah and Abraham, for example—before the corruption in the world and the wilfulness of men had led to the establishment by Moses of a system that catered for a more imperfect society. Moses had imposed the religion of the Law which acted as a kind of bridle, a disciplinary regime; it was now regarded as a thing only meant to be temporary. This view that the primeval purity of the faith was now brought back, but with an added dimension and in a transcendant form, became the key to a lot of things. It meant that Christians could use world history in a way that gave them a faith not only against the Jews but also against the pagans. If the ancient Hebrews had so declined from the primeval state that they needed to be put into the straitjacket of Mosaic law, at least they had kept monotheism alive, kept the religion of Jehovah in existence. The rest of the primitive world, the pagans, had declined further still, and had merely sunk into polytheism.

All this represents the first stage in the production of a scheme of world-history in which Christianity was to be related to everything that had gone before it, put into its place in the whole time-series. The very fact that it bridged both the Gentile and the Jewish worlds meant that Christianity was able to have conceptions, forced to have conceptions, about the mundane history of all mankind. The Scriptures themselves contributed in a powerful manner to such a view, since they started with the Creation, told about the primitive states of the human face, and went on to describe how the world had come to be divided into so many nations and languages. The controversies forced the Christians to turn their attention to the more distant past, and to think out the position in which they stood in the march of ages. Once their thought came to be turned in this direction they could not help reverting to the view that the Old Testament was to a considerable degree a history book, the basis of the very kind of history that they required. In fact, the book of Genesis, besides inspiring the Christians with the notion of what we might call global history, set the pattern for the opening section of all universal histories until the eighteenth century. It has generally been the case, and it has remained substantially the case, that the men who write political history set out to tell the history of their own state, their own city, their own nation. It is from religious or quasi-religious ideas that there springs the desire to see the human race as a whole and to envisage universal history. And, starting in the manner that I have described, Christianity gave a great impetus to this notion of universal history.

Even so, its achievements were reached by a curious route. The Christians were under the necessity of answering not merely the squalid polytheism of barbarian peoples who indulged in cruel practices, but the high and noble kind of paganism which was exemplified in some distinguished figures in the high civilisation of the Roman Empire. As the Church spread into the more educated classes it came to have leaders who greatly admired the loftier sections of Greek philosophy, work which had pointed to an ultimate monotheism. Some of those who became Christians did not lose their love of Plato, for example, and they tended to expound the religion itself in terms of Greek thought, the highest philosophy that they knew. Christian thought in the time of the Roman Empire, therefore, had in a sense a double root—it went back not only to ancient Judaism but also to ancient Greece, and it soon claimed to have captured the best of both these traditions. It proved possible to show, however, that the wisdom of the ancient

Hebrews was older than that of the ancient Greeks. The earliest philosophers amongst the latter, and even Homer himself, were shown to be nothing like so old as Moses. Even Pythagoras and Plato were shown to have come later in time than some of the greater prophets of Israel. It came to be held that Plato himself, in the higher region of his thought, did no more than restate what he had learned in one way or another from the ancient Hebrews. And he, like the men of other nations, had even perverted part of the wisdom that he had borrowed. In other words, the ancient Hebrews, who had so long been neglected by the Graeco-Roman civilisation, were now presented as the oldest people of all, the key to the history of culture, and the source of such wisdom as had spread amongst other nations. Their language was the original language of mankind, before the confusion of tongues took place. It was the language of God. And, like the Jewish historian, Josephus, who lived not long after the time of Christ, the Christians reversed the original charge against them, and reproached the Greeks with being mere newcomers, mere children, only appearing as a people quite late in the day. At the same time, the view that the Mosaic system, the religion of the Law, was only a provisional dispensation, intended for the training and the disciplining of the Jews, until, in the fulness of time, the Christian revelation should appear, involved the notion that God had a plan, a scheme of salvation which needed time for its fulfilment, that time itself was a factor in the story, and that after the lapse of centuries men would be fit for something higher. By about the year 180 A.D. such a notion was becoming explicit in the work of St. Irenaeus, *Against the Heretics*, against the Gnostics, that is to say.

But before you could prove that Moses was prior to the Greek philosophers you had to tackle the question of the chronology of world history. This was quite an elaborate enterprise because so many nations and peoples, even separate cities, had had different chronological systems, different ways of expressing the date. It required quite a science to discover how all these various calendars fitted into one another. Before it could be decided that Moses was anterior to the Trojan War you had to find a way of synchronising events under one calendar with those under another. You had to work out what monarchs in various countries were living at the same time. The Christians—forced to consider both the Old Testament history and the story of the Gentile world—developed in a remarkable way the attempt to find a universal chronology. Because it entailed correlating events in one region with events in a quite different region, the enterprise did actually involve them in universal history itself. The Old Testament chronology was the key to the system adopted by the Christians, of course, and, once again, we see the ancient Hebrew Scriptures asserting themselves as sheer historical record—for Christians the most authoritative of the sources for the ancient world. At the same time, the interest in universal history made it necessary to dovetail the pagan historical work into the framework provided by the Jewish Scriptures. Some time not far from 221 A.D. Julius Africanus elaborated a whole chronology of world history in what was to prove an important pioneering work. It is not surprising that this led him at the next stage to the production of a sort of synopsis of world history.

The attempt to see all the past as a single great drama, involving the whole of mankind, had a curious result which left its mark on the writing of universal history until the beginning of the eighteenth century. One of the characteristics of Judaism in the latest centuries before Christ had been the tendency to periodise history—to divide it into epochs—and to see each epoch as having its own character, as being under a dominating force. It was a practice stimulated by eschatological speculations, the descriptions of the successive stages in the working out of the Last Things. In a sense it was possibly a division outside ordinary history, though it still represented an effective division into separate eras, separate areas of measured time. I think there can be no

doubt that to the people of Israel, the rise of colossal neighbouring empires, which had begun to make it impossible for them to
have an independent political life, had come to be interpreted in
a new way. It had come to be seen as the beginning of the end.
The first catastrophe—the famous Exile, the dispersion to Babylon—had been taken as a judgment of God in the normal way.
But as one vast empire succeeded another, making life impossible
for a small people, a small state, this phenomenon of world empires began to assume a demonic character and a cosmic significance. The Jews caught from abroad the theory of the Four
Monarchies, the Four World Empires, which makes its appearance in the book of *Daniel*. The theory became the basis for their
periodisation of world history. And the Christians took over this
idea; again the Old Testament was a decisive factor in their shap-
ing of world history. By the Christian era the Four Empires
tended to be interpreted as being the Assyrian and Babylonian,
the Persian, the one founded by Alexander the Great, and finally
that of Rome. It was still held that there were only to be four of
them, and they stood as the last things in human history, a kind
of prelude to what would really be the end. In the Judaism of the
first century A.D., and in Christianity rather later, it is explicitly
held that the continuance of the Roman Empire is the only thing
that is holding back the End of the World. And it is interesting
to see that Tertullian, early in the third century A.D., says that
Christians want to defer the end, and that, partly for this reason,
they support the Roman Empire. This theory of the Four Empires was later to become the accepted Christian basis for the periodising of world history. It received a new lease of life at the
Reformation. It continued to be used into the seventeenth and
even in the eighteenth century.

Even after Christians had ceased to be holding themselves tense
in the expectation of the absolutely imminent end of the world,
they continued to speculate about the possible date of the end, in
spite of the fact that Christ could be quoted as saying that it was

not for them to know the times and seasons. And these speculations for a long time seem to have interested them much more
than actual history did. In the *Epistle of Barnabas*, which may go
back to between 70 and 130 A.D., it was argued that the Creation
of the world had taken six days. The habit of taking in the Old
Testament text as not merely a record of what had happened but
a prefiguring of the future led to the conclusion that the world
itself would last for six corresponding "days," though each of
these "days" really meant a millennium, for in the *Second Epistle
of St. Peter* it was written that "one day is a thousand years with
God." Christians accepted the biblical chronology, and swept
aside the old Babylonian and Egyptian teaching to the effect that
the world had already lasted far longer than the full 6000 years.
On the biblical chronology it was estimated that Jesus Christ was
born between 5000 and 5500 years after the Creation. The end
of the world was only a few hundred years away, therefore, and
what we call millenarian speculation ran riot.

II

We should remember, in considering the next stage in the story,
that people in those days envisaged a remarkably small world, to
us something like a toy world, with the stars overhead as part of
the scenic background, and the sun specially created to be of
service to men. Amongst the Jews one finds the saying that Jerusalem was the centre of the world, that God created Jerusalem
first and then the rest of the world around it. Dante was to have
a similar idea of the extent of the land on the globe, and he, too,
saw Jerusalem at its very centre. It was easy to insert a sort of
symbolism into a geographical point like this, and indeed into certain things in history itself. There is a similar symbolism in Aristotelian physics where the noblest things—fire and air—rose to the
top and the heavenly bodies themselves were made of an ethereal

kind of matter. The time-scale was small, and now the world was coming to an end. Jewish writings of the first century A.D. say that the world was getting old, nature herself was becoming exhausted. It was all a small affair and the earth was the scene of a small-scale human drama, of which the final act was now supposed to have begun. For both Christians and non-Christians the air was full of active spirits, some of them wicked demons.

The moment arrived at the end of the third century A.D. when we can say that historical consciousness had developed. It found expression in Eusebius, a leading personage who had no use for millenarian speculations. He had a mind for more concrete things, the things that had actually happened. The moment had come, moreover, when Christians, looking back, could take stock of their whole enterprise, assessing their history as though it was a chapter of completed story. Perhaps it was a deceptive moment for churchmen to choose for taking their bearings and working out an interpretation of history. One of the reasons why they became history-conscious at this time was the fact that they themselves saw events come to a great climax. They could feel that a tremendous kind of history-making was going on around them. By this time the Church comprised a very considerable section of every province in the Roman Empire. It suffered the worst and most bitter of the persecutions, but it surmounted these and, through the conversion of Constantine, it captured the Roman government. This is the high spot in the story of the rise of Christianity. At this point there emerges Eusebius, developing his ideas in a number of learned works in the decades before and after the year 300 A.D.

In general, he set out to meet the charges that Christianity appealed only to the feelings of ignorant men and that it was a newly invented faith. He tried to show how it had recovered the natural religion of primeval man, and how it had collected into itself the best that ever existed in either pagan or Hebrew history. In his view the Hebrews were the real sources of culture, the first people even to study the material universe. The only people, he said, who from the first devoted its mind to rational speculation. He followed Josephus in taunting the Greeks as being mere children, born only the other day. They had plagiarised all their philosophic lore, he said, and they had added nothing of their own, except force and elegance of language.

According to Eusebius, Christianity owed its chief debt to the Hebrews. From the Old Testament it took over the Prophecy not the Law. In this connection, he maintained that the prophecies contained hidden secrets—"disguised," he said, because the Jews would have destroyed the writings if the predictions of their doom had been written plain. Eusebius did not consider Moses so important to the Christians, since his work was only of local and temporary importance, limited entirely to the Jewish people, not practicable even for the Jews of the Dispersion, not practicable for anybody who did not live in the Holy Land. The system of Moses was a narrow and interim affair. It was like having a doctor to heal the illness, the demoralisation that set in, after the Jews had been contaminated by contact with the Egyptians. The Law was "like a nurse and governess of childish and imperfect souls." There is almost a hint of progress in this view, a glimpse of the idea one catches in Irenaeus, the idea of the "pedagogic" function of the time. It is perhaps significant that Eusebius, writing on this subject, calls his work "The Preparation of the Gospel."

It is when Christ appears that mundane history in the writings of Eusebius seems to take an almost magical turn. It is not sufficient that Christ comes in the fulness of time—He comes, in accordance with prophecy just at the moment when the Jews have no king of their own line. In the book of *Genesis* (49:10) it was said, "There shall not be wanting a king from Judah or a leader from his loins until he comes for whom it is reserved." This is now taken as being a reference to the Incarnation which occurs when the Roman Emperor Augustus imposed on the Jews a foreign monarch, Herod. Christ coincided also with the establishment of a Roman Empire, comprising the bulk of mankind, and producing peace, facilitating communications, over a wide area—a providential arrangement for the preaching of Christianity. Furthermore, soon after the Crucifixion, the Jews rebelled against the Roman Empire. Jerusalem and the Temple were destroyed. The Jewish people suffered a final dispersion, the final punishment which came very quickly after their rejection of Christ. There were other remarkable happenings. From the time of Christ there was an improvement in the habits and customs of even the pagan world, we are told. The heart began to go out of pagans from the mere fact that Christ was in the world, before there had been any actual preaching to the Gentiles. From just this moment the oracles ceased to function, and human sacrifice finally disappeared. Above all, Christ had won a victory on the cosmic level—he had thwarted the evil demons. If they fought desperately still, it was because they realised that their doom was now really decided. This meant that there was a general drift in the world at large, a general drift away from polytheism, just when Christianity was due to come in to take its place. It helped the establishment of peace, for the multitude of the gods had been connected with the multiplicity of the nations, and so had been responsible for the repeated wars.

From this point in the story, Eusebius becomes virtually the founder of what we call "ecclesiastical history." In this aspect of his work one of his objects was to establish the succession of bishops in the important sees—a momentous matter, since authority had to be traced back to the original disciples. He also desired to commemorate the martyrs—a theme to which he was particularly attached—and to leave a record of what they had suffered during his own lifetime. Then, again, he wanted to give an account of the successive heresies, though here he failed to enter into the historical point of view, failed to explain the case for the other party or the reason why the troubles arose. He was too convinced that heresies were the work of wicked demons. On the other hand, when he comes to his own time, he uses very strong language about the evils that had arisen in the Church in a time of considerable prosperity. He regards the persecution, in part, as a chastening permitted by God for this reason.

It was the culmination of the story by the time of Eusebius himself which was so remarkable. The policy of persecution had to be abandoned. The most successful emperor of the time, the famous Constantine, was a convert to Christianity. By his own account he achieved military victory through a miracle, and found that the Christian God was the one who succeeded in battle. By his own account he received direct messages from Christ. Eusebius tells us how, by divine means, he learned the devices of his enemies in advance, gained the foreknowledge of future events, found the expedients to employ in times of extremity, and even came to his military arrangements. The Roman Empire had been a glorious thing to Eusebius, but now it was to be a Christian Empire too. Now it was drawn together again by Constantine, the eastern half reunited with the western half. Constantine "alone of all rulers" had pursued an unbroken career of conquest. He had gained authority over more nations than anybody before him. He was the first Emperor since Augustus to celebrate three decennials, to reign for over thirty years. And Eusebius, who if he liked to record the sufferings of the martyrs and liked to note the early and violent deaths of persecutors, was able to point out that Constantine, in contrast with many recent emperors, was allowed to live to a good ripe age.

In Eusebius' view, from the beginning of time there unfolds a grand design of Providence. Everything is beautifully patterned and symmetrical, the past containing symbols and prophecies which pointed to the glorious future now becoming an actuality. The story was approaching its final consummation. There were only a few nations outside the system, only a few nations to be gathered in. In a sense it was a glorification of sheer success—success which indeed might have a certain spiritual aspect, but was also of a very tangible mundane kind. Eusebius jeers at the ancient gods who are going into decline, of course, and are proving so incapable of doing anything to stop the rot. He jeers at the oracles which failed to give warning of the catastrophe, failed to give warning of the advent of Christ Who was to put them out of business. For Constantine himself the Christian God is often the most effective of the wonder-workers and magicians. God is really the One who sees that his followers prevail in actual battle. Eusebius sometimes produces the same kind of impression when he personally tells Constantine to impute his successes, even his military victories, to God. I am interested in the way in which Eusebius, even when he is producing a congratulatory oration in the presence of Constantine, will talk in fact much more about Christ than about the Emperor. He wants Constantine to connect Christ with the successes that he had achieved in the world. It is wrong to regard Eusebius' view as too much an effort to gain imperial favour; more is involved. In his admiration for the Roman Empire, especially after its government had become Christian, he produces a kind of political theology, and even secular history seems to become sacred history in his hands, and the rapprochement between religion and worldly success seems over-crude. There are subtle ways in which the affairs of the spirit can become entangled with the affairs of the world, and Eusebius, who in any case would overlook the paradox of the wicked emperors who were allowed to enjoy a long life, allowed his assessments sometimes to be based on externals. He would have had to make a different picture of the ways of Providence if he had been taking his hearings a hundred years earlier or later. Though he is more of a historian than any of the other leaders of the early Church, he is not the one to whom we should go for the real source of a Christian interpretation of history.

III

Eusebius belonged to a time when Christians could hardly help feeling that human history had come to another tremendous climax. Persecution had been surmounted; the mighty Roman Empire had come under Christian leadership; one of the greatest of the emperors, Constantine, had converted. St. Augustine, a century later, confronted a different kind of scene and brought to the examination of history a different kind of experience. He had seen something of the evils that could exist even when the world was under Christian rulers. In his day, that Roman Empire, which many had deemed eternal, was being overrun by barbarian peoples. In 410 A.D. the city of Rome itself was for a short time in the hands of these barbarians. He had to answer the charge that the disasters were due to the neglect of the ancient gods and the triumph of the Christian faith.

Augustine met his particular problem by producing at a lofty level a significant kind of historical study—a very ambitious analysis of what I should call the whole drama of human life in time. He wrote as a believing Christian and regarded the inspired Scriptures as containing the most accurate historical writing in existence. Unlike Eusebius, he did not love the collection of mere facts, the hunt for sources or the recording of contemporary events for the sake of the future. He accepted the data provided in the stock classical histories or in the Old Testament, though he could do a bit of ingenious criticism when faced by a glaring anomaly. He was even perfunctory when he had to do any narrating, any mere recapitulation of monarchs or wars or famous occurrences. It was part of his plan to show that Rome had suffered no end of disasters before the Christian religion had even appeared. He had not the patience for such a work of enumeration, and he handed that part of the task to a disciple of his, Orosius, who worked the matter out in a book of his own.

He was not primarily concerned with what we today would call straight history. His subject was the whole human drama, the spectacle of this human race stretched out across all the ages. He is forever turning aside to discuss the fundamental problems of human destiny. Why and how did the world begin? Is the human race merely the prisoner of a kind of fatality? How are we to deal with the problem of human suffering? Some of his questions, however, come closer to being historical in our sense of the word. Why are primeval men recorded as having had longer life and greater stature than we should have thought possible? Those men who lived for hundreds of years, at what age would they begin producing children? Where did civilisation take its rise? How ancient is the wisdom of the Greeks? Why were the primitive Romans so successful and how did their successors come to establish an empire so extensive, so enduring? He had a remarkable way of handling the problem cases, the discrepancies, for example, between the Hebrew Scriptures and the ancient Greek translation of them, the Septuagint. And taken altogether his book must remain—outside the ancient Scriptures—the supreme example for study if one is interested in the connection between history and belief. Indeed, he is one of the very greatest minds that ever set out to discuss the human condition, and so to tackle history at a really fundamental level. The work of his which concerns us, *City of God*, presents us with a paradox. In a sense he is nearer heaven, much more spiritually profound, than Eusebius, but to a twentieth-century historian he is also nearer to the earth, with a very much better idea of the way in which history works.

His superiority, especially in relation to Eusebius, is shown almost at the start in the way in which he tackles the problem of God's judgments in history. It is as though he were trying consciously to reverse a good deal of Eusebius and do a little demythologising of his own. He insists that God sends the sunshine, the rain, and the blessings of the world on the good and the wicked indifferently. This applies, he says, to the acquisition of thrones and empires and to success in war, things in which divine Providence is most emphatically involved. It applies even to the gift of long life. St. Augustine will have nothing to do with the view hitherto current that God gives the Christian kings lengthy reigns but sends the persecuting monarchs to a quick and terrible death. If God rewarded the Christians with mundane happiness, they would think too much of worldly prosperity and regard it as a sign of spiritual grace; you would have men becoming Christians for the wrong reasons. The good differ from the wicked in the way that they take, the way that they use, misfortune. They regard it as part of the discipline of mundane life, and as a thing which tests their virtues and corrects their imperfections. Christians will accept even misfortune itself as a judgment or a chastening or a reminder of their own shortcomings, for they, too, have their part in man's universal sin. The best of them will have some lurking infirmity, perhaps pride in his own righteousness or anxiety about his repute in the world. None of them can claim to be suffering as absolute innocents.

Yet this same God, according to St. Augustine, does actually give mundane rewards to men, rewards even to pagan virtues. Augustine sets out to examine the secret of purely worldly success, and the crucial example for him is the rise of ancient Rome. He will not allow that the heathen deities gave their assistance here—it was a matter only within the Providence of the Christian God. And he writes, "Let us consider what were the virtues which the true God, in whose power are also the kingdoms of the earth, chose to assist, in order to produce the Roman Empire." In that primitive city of Rome there were pagans who were ready to sacrifice their private fortunes for the good of the state, prepared to brave death rather than sacrifice liberty. Even if after that they went on to seek domination over others, they did it for the love of glory; it was better than living for wine, women, and song. These men were "good in their own way," says Augustine; they were "laudable doubtless and glorious according to human judgment." They knew what they wanted. They wanted power in the mundane sphere, and they disciplined themselves and sacrificed their private comforts to secure it. They were so faithful to the earthly city that they ought to stand as an example to those who are members of the heavenly city. They could not be candidates for Heaven, of course, but, says St. Augustine, if God had also withheld terrestrial success from them they would not have received the reward for their virtues, the terrestrial reward that was due to them. His argument culminates in the quotation: "Verily, they have their reward."

In spite of his overwhelming preoccupation with the things of the spirit, he recognises the existence of profane history, and he came near to conceding to it a certain autonomy, though not quite. He can insert mundane ideas about causation. He says that the recent despoiling of Rome, after the eruption of the barbarians, was the result of the customs of war. He accepts the earlier, pagan view that the destruction of Carthage—the elimination of the one thing that Rome had to fear—had, by ridding the citizens of anxiety, produced a relaxation of discipline and morals, a decline of public spirit. When the expanding republic came to be harassed by social and civil wars, he did not say that this was a judgment of God, though his doctrines might well have allowed him to do so. He said that Roman conquests had become too vast and her empire was breaking under its own weight. So far as the field of profane history was concerned he had a more flexible view of the workings of Providence than Eusebius. He does not pretend to know God's purposes in profane history so clearly, even though in regard to sacred history, salvation history—in regard to the Incarnation, for example—he would have seen events taking place according to God's prearranged plan. The God who gave vast empire to the great Augustus, gave it also to the cruel Nero, because it was the government which the people of the time deserved, he says. Augustine is not so ready to see the Roman Empire installed by a kind of magic to coincide with the coming of Christ. He recognises that the Roman Empire united many nations and created a great area of peace in the world. What he stresses is the fact that Christians share in the blessings of peace that the secular state provides. They owe something to the body politic which enables them to have the material necessities of life. Judges in the terrestrial city sometimes torture innocent people, and cannot be sure that they may not have condemned an innocent man to death. Still the office is important, and wise men ought not to reject the responsibility of it. The Christian will realise that he can never be happy in it and will never long for it. He will rather pray that he will not have to drink this cup, but if he must accept the office, this is the spirit in which he will. And the same is true even with the office of the Emperor himself—the Christian will accept it as a means of service.

All this Augustine concedes to profane society and discusses almost in the terms of a secular historian. Yet he must also have had a deep hostility to Rome, which we find him covering with vituperation and treating as the second Babylon. We find him still more remote from the political theology of Eusebius, with its sanctification of the Roman Empire, when he declares his preference for a world of little states living in unity rather as families live side by side in a city. He is aware of the human cupidities which make that ideal so difficult to achieve, however, and he seems to accept the fact that it was the turbulence of the neighbouring nations which goaded Rome into fighting them and conquering them. All the same, he has an exceptional hatred of war. When people talk of the *Pax Romana*—the vast area that has been pacified by Rome—he always wants them to remind themselves of the terrible bloodshed through which that peace had been achieved.

In any case, the virtues of the profane world and the importance of profane history are only relative to him. Rome belongs to the earthly city and the state is a combination of people for the procuring of mundane ends. On a spiritual view even the relative virtues become vices, because the motive of such combination ought to be the love of the heavenly city. Yet the profane is not evil in itself. God did not create anything evil, and even the devil is not evil by nature. He almost says that evil itself is not absolutely evil—God would not have allowed it to exist if he had not foreseen some good that he would be able to achieve by it later. The darker side of life is like the shadows in a painting—the contrast enhances the beauty of the whole, though we who are entangled in these things are not in a position to see it. Things themselves are not evil; only the excessive love of them is. The love of profane things is not evil; the sin occurs when men devote to the lower things the love that they should give to the higher. In other words, the evil lies not in things themselves, but in the will of man which chooses the lower rather than the higher. Here in the actions and choices of the will—things not determined by God— you have the real evil, evil *par excellence*. For this reason, when he is wearing one thinking-cap, when he is speaking as a profane historian, he can talk of the virtues of the ancient pagans. When he is wearing another thinking-cap, those virtues are themselves described as vices. He gives profane history a place, though its virtues are only relative.

He differs from his former Greek philosophic teachers in that he recognises the Christian's commitment to history. It is rather a thrilling thing to be able to see how, in his *City of God*, he argues his way out of the Greek cyclic view of the process of things in time, even the extreme version of it which asserted that all history goes on exactly repeating itself throughout endless ages, everything happening over again in the same way. Any form of the cyclic view of the time-process turning its aimless revolvings and pointless repetitions robs history of any significance. Augustine saw that the idea of Christ returning to be crucified again in another repetition of the cycle would turn the whole salvation story into a kind of cosmic puppet show. The eternal bliss that was to be granted to the saints was utterly inconsistent with the idea of everybody having to return and to repeat the miseries of this mortal life. It has been plausibly suggested that his attitude was affected by the pull of the Old Testament, which he regarded not merely as prophecy and symbol but as actual history. He certainly picked up the Old Testament view that history is really going somewhere, or at least pointing to an end, when God should have completed the number of the elect. For this reason Augustine saw that there is some meaning in history, and in a certain sense *City of God* is his attempt to work out this meaning.

IV

I have mentioned that Augustine confided to a younger disciple of his, Orosius, the task of showing that Rome and the rest of the world suffered plenty of evils before Christianity had ever appeared on the scene. This man, though he compiled his work from writers of various nations already current and popular at the time, had more interest in the concrete data than Augustine. It is his work which we should describe as the first real Christian attempt at universal history. There was a sense in which he took a universal attitude. He did not write as a Roman looking at the story of mankind; he was rather a hostile witness, trying to show the evils of pagan Rome. He was careful to point out how the glories and the happiness of conquering Rome meant misery for all the other peoples whom Rome had conquered. It was partly his case that there was hardly any cessation of war and massacre and cataclysmic disaster until the time of the Incarnation. In fact, his was a universal history definitely commissioned from him, definitely produced by him, to show the incessant miseries of mankind throughout almost the whole of time. He reinforced his case by some very interesting exercises in what we should call "historical explanation," as, for example, a number of expositions of the way in which men fail to make history vivid for themselves because the memory of the biggest disasters of the past, the mere reading about colossal massacres long ago, can never produce anything like the anguish that even a little pain—even a sting from a fly—can give us in the actual present. And, coming from Spain, a region which had struggled bitterly to escape domination by the Romans, he was able to see Rome rather from the position of an outsider. Indeed, writing for Romans, he had to expound for them the necessity for using the imagination in order to learn how the other party might envisage a given event, how, for example, suffering Carthaginians might have felt, so that he was beginning to realise something of what was necessary for achieving what we should call a historical point. He shows historical sense, for example, in the way he foresees that the present barbarian invaders might conceivably settle down and form an order in a culture that will be as acceptable to the world as the Roman one. At the same time he did not enter into the whole of the thought of St. Augustine, and in some respects his work represents a regression to something more like the view of Eusebius. He fell very much into the view that God rewards piety with worldly success, that persecuting emperors meet early and terrible deaths, while Christian emperors win easy military victories. He carries still further the view that the coming of Christ changed the character of the world itself. Even before the preaching of the Gospel and the conversion of the Gentiles ,it produced a moral improvement in paganism itself, and an era of peace. Henceforward the wars and miseries of mankind were fewer. But not only this. The cataclysms in nature were themselves less terrible. Even the volcano, Mt. Etna, no longer boiled over in fury. It merely emitted an innocuous cloud of smoke, once the Christian era had opened.

All this was to be significant, because it was Orosius rather than Augustine who influenced the succeeding centuries. The middle ages, when they thought they were following Augustine, were almost invariably following Orosius in fact. Orosius was more easily understood. His work became one of the really influential books of world history, and the number of manuscripts, editions, translations through which it passed is colossal. Until well into early modern times it was the work of Orosius which really set the pattern for the study and interpretation of world history.

8

Does Belief in Christianity Validly Affect the Modern Historian?

I

We are sometimes faced by the question: How far does the fact of being a Christian believer affect—or how far *ought* it to affect—the work of anyone who engages in the writing of history, or undertakes researches into the past, or attempts the interpretation of public events? In the discussion which such a problem entails, a certain logical priority ought perhaps to be given to the obvious duty of the historian to establish the hard facts, just discovering the things that actually happened in the past, and seeing what can be properly deduced from the tangible evidence that has come down to us. And this must be regarded as grounded in a technique, a field in which even clever and well-meaning people can go wrong if they do not know the rules and have not been properly drilled in such things as criticism and the use of evidence. Indeed there have been natural scientists—men who would never have made a mistake in their own kind of technical studies—who have sometimes come wide of the mark when writing about the history of science, because they have imagined that the handling of historical evidence involved nothing more than ordinary common sense. Furthermore, once the historian has established the hard facts, the external events—the things which, in one way or another have managed to leave their mark on the tangible evidence which has survived—he then moves forward to what may be regarded as an associated achievement or even a higher task: the demonstration of the connections and interlockings between one event and another. And, in many respects, this further aspect of the historian's function will need to be carried out, too, in a more or less scientific manner.

Now I, personally, would never regard a thing as "historically established"—that is to say, as genuinely demonstrated by historical evidence—unless the case for it could be made out in a coercive and inescapable manner to any student of the past—Protestant or Catholic, Christian or non-Christian, Frenchman or Englishman, and Whig or Tory. Furthermore, the historian must play the game according to the rules. Within the scholarly realm that is here in question he is not allowed to bring God into the argument, or to pretend to use him as a witness, any more than a scientist, examining a blade of grass under a microscope, is allowed to bring God into his explanations of the growth or decay of plants. In the middle of the 1930's half a dozen history examiners conducting a *viva voce* examination in Oxford were sorely tried by a candidate not by any means stupid or unattractive who insisted that God—and only God—had been responsible for the French Revolution. It proved impossible to induce the man to look for any more immediate origin or explanation for the cataclysm; nor could he be persuaded to give his mind for a moment to the consideration of what used to be described as "secondary causes." In my university teaching during the same period I would sometimes choose to ask a student to explain the victory of Christianity in the great conflict of religions that occurred over the length and breadth of the ancient Roman Empire, and I would decline to accept the answer that the success of Christianity must be due to the will of God, or the work of Providence, or the fact that the right religion was bound to prevail. Like Gibbon, I intended my question to mean, what were the secondary causes, the reasons under God, for the victory of Christianity? In other words, I was really enquiring about those factors in the mundane realm which helped to give Christianity its extraordinary survival value in those ancient days. And the historian seeks the historical explanation of what happened, just as the physicist will give a scientific explanation of what has happened in his laboratory. Both historian and scientist offer a partial explanation—each in his own terms—but not by any means a total explanation of anything.

In any case, it must be confessed that not by any means will all—perhaps indeed only a comparative few—of the things asserted in an historical narrative have proved capable of being demonstrated in an absolutely coercive manner, even where their existence may be regarded, for ordinary purposes, as sufficiently probable. It is necessary to remember how unreliable, in one way and another, all eyewitnesses, however honest, and all reporting, however clearly it may seem to come straight from the horse's mouth, can be. In any case, not only may later information emerge to correct at unexpected points the information now regarded as definitely established, but the whole context around some particular event may turn out to have been falsified, and some changes in the understanding of human beings in general may alter the bearings of events which in themselves had been correctly established. The thousands of history research students who are working in archives and libraries at the present day would come to feel that their work was a waste of time if they failed to produce advances in knowledge and changes of interpretation, even in fields where the general course of things had seemed absolutely settled. And the changes that the passage of time tends to produce in the situation of affairs in one region or another may subtly alter even the interpretation of the literary sources of the past, or the judgment that we make on bygone generations. Historians, especially when they are dealing with merely incidental points, have often made use of memoirs, private correspondence, and miscellaneous forms of mere "reporting" which for a considerable period they will subject to only routine and haphazard criticism, the new evidence being envisaged for some time in the framework of story already accepted by contemporaries. It is well, therefore, if historians can be persuaded not to hold the shape of past events too rigidly in their minds. And, in any case, it should be remembered that the business of establishing a negative in history is more difficult than is often realised, not to be settled finally, for example, by a mere absence of evidence.

If all this is true, it might be dangerous for a historian to be a Christian, dangerous at least for him to be too much in earnest about his Christianity. Indeed, when one thinks of all that has been presented to the public during the last two thousand years and more, one might be tempted to wonder whether anybody has perverted history more than Christians have done, on this occasion and that, whether for the sake of their religion or for the service of a particular sect, or to establish the glory of a favourite saint. And this can be the case, even though it may be true on the other hand that religion itself has provided the motive for discovering and circulating highly inconvenient truths, and may have had much to do with the existence of the notion that there exists a truth which makes a supreme claim on human beings. It used to be debated whether the life of Christ, or the history of Christianity, would be best produced by an actual follower of the faith or might reach a more judicial result when written by a non-believer. But an author who in the former case might be too much under the domination of his theological prejudices might, in the latter case, lack the proper and requisite feeling for the religion involved, a religion of which he might never have had any considerable personal experience. And, though somebody has suggested that the ideal person for such a task, therefore, would be a man who at one time had been a Christian but at a later date had thrown his religion over, this thesis is itself open to the retort that a man who ended by deserting his faith might never have had a very adequate experience of it, never had a realisation of its inner quality. In any case, if the more religious ages of history have shown sometimes the unfortunate effects of theological sympathy or theological prejudice on historical writing, the more thoroughly secularised modern world has quickly made it plain that men without the slightest belief in anything supernatural can go just as far as the saints in perverting historical narratives and persecuting one another. In fact, it would appear that there are always insidious dangers for the historian, whether the man in question is any sort of Christian or any sort of non-Christian. In such a situation, it is obviously best to be self-denying in a sense, and to recognise the unavoidable constrictions, the subtle dangers, in the various techniques connected with historical study. And it might be urged to Christian believers that, for them, an absolutely necessary concomitant of historical study is intellectual humility, together with a due respect for the philosophies that are not one's own.

The historian who is at the same time a Christian and who may be stimulated in the first place, but may be seriously limited for the rest, by the tangible, external evidence that he can accumulate, will hardly feel permitted to consider the drama of human history without picking up also the opposite end of the stick, and remembering the significance of the internal life of human beings. When each of us at the present day looks at himself, he is aware of something which makes him feel solitary and unique. He seems to gaze upon the world as from an inner citadel, and he is sensible of exerting action upon the things he sees, building up a fabric of personal experience and having his own realm of inner thought. Each of us comes to infer—though perhaps never actually observe—how the people around us seem to have similar feelings to ours, or similar reactions in the face of vicissitudes; though in the case of other people our apprehension of this inner life of theirs is less clear and accurate and assured, lacking the immediacy, the directness, of those things that happen inside ourselves. The world is so constituted, indeed, that American citizens who are terrified by what they feel of the appalling power of Russia may be unable to comprehend, unable even to credit, an exactly parallel counter-fear which their own military strength, or their own political conduct, may have been responsible for provoking in the Russians themselves. And Americans who are totally—and rightly perhaps—assured of the sincerity of their own good intentions, may find it impossible to have a similar confidence in the Russians, since, however virtuous these latter may be, it is never possible for an outsider really to know the other person from the inside. Indeed there is perhaps no point at which historians have shown themselves so curiously lacking in sympathetic imagination and human understanding as in those cases—the case of Germany in 1914 for example—where the "other party," the potential enemy, the dreaded competitor, has claimed that some offence of his has itself been prompted by fear.

The student of the past has to treat historical personages as we ourselves have to treat the people around us. He, too, has to work from outside inwards, making the best of the various kinds of external evidence that have come down to us. Some things will be easy for him. He may, for example, feel clear that at one moment Napoleon was unhappy; at another moment he might say that Napoleon was angry. But he may have to remind himself that a favourite device of this man was—for tactical reasons—to give just the appearance of being in a temper. When the historical personage seems to be giving the most intimate piece of personal confession, the student may have to enquire how far even this may have been written with an unavowed tactical intent. Indeed, if we wish to see another personality from the inside—really following his feelings and his processes of thought—we must surely go to the novelist, and even he can only accomplish the task for personalities that he has created himself. Moreover, even the long record of a single day in James Joyce's *Ulysses* is hardly sufficient to provide all that the student is likely to need. History is quite wonderful for the way in which, when a man has taken a certain decision, it can reassemble so many of the considerations that were in his mind, or might have been in his mind—so many of the factors of which even he himself had no clear consciousness—before taking the crucial step. But it is much more difficult—and surely often impossible—for the historian to seize upon the particular consideration in the stateman's mind which finally tipped the decision in one way or another. History, in fact, can show that the options open to a statesman are at any rate always fewer than are imagined by young people conducting politics from an armchair. But it can never know that it has reproduced the whole complicated fabric of conditioning circumstance—for, after all, a man's conduct at a given crisis may even be affected by something that he had recently had for dinner. I doubt whether one's materials enable one to say as an historian that, given the same enormously complicated network of causal factors, one would have made a better use of one's chances and materials than some of the people one so badly wants to criticise. Nor do they enable one to say how far a man is himself really responsible for being the kind of person he is. *A fortiori*, the historian as such cannot pronounce upon the divinity or the nature of Christ, cannot indeed make this ultimate kind of judgment on the inner nature of any personality, human or divine—though, if he could prove that nobody had ever suffered crucifixion under Pontius Pilate, he would provoke some comment on a religion which for two thousand years had been regarded as hinging on this very point.

Historians do not remain mere technicians, therefore, and they must bring all that they know about life and themselves, all of their personal experience and their accumulated knowledge of the centuries behind them, all that they have ever felt in the deepest parts of themselves, to enrich their understanding of human events and even their interpretation of pieces of evidence. The

Christian historian will certainly have a place for that humanistic sort of narrative that governed the treatment of the past for so many ages. He will remember that, as far as concerns the ordinary mundane realm, it is men who make history, each of them standing as in a certain sense an end in himself. And in any case the story must not be reduced to mere process, a mere case of watching the interactions within a piece of machinery. Indeed, even while working as a technical historian, the Christian will find himself committed to the view that human beings have deep interiors, complicated corridors, and tremendous caverns inside them, and he will have the feeling that there is something more than becomes explicit in the evidence, something so subtle that it escapes the historian's fishing net. And this might soften what would otherwise be the excessively hard lines of the historian's diagram, and might even double that flexibility of mind which is so important for him and which he so needs to possess.

II

The study of history in our quarter of the globe can be traced back to ancient Mesopotamia in the two or more millennia that preceded the birth of Jesus Christ. Throughout this area and throughout these ages, religion played a great part in the narration of events and in the meditation upon the past—it governed the celebration of victories, the thanksgiving for harvests, and the mourning of catastrophes, for example. And, centuries before the time of our Old Testament, the misfortunes of the body politic were habitually associated with offences committed against the gods. The historiography of the region was brought to a further stage by local developments taking place further west, somewhat nearer to the eastern end of the Mediterranean Sea. Here the ancient Hebrews could not forget how God had rescued them from slavery in Egypt, and, through many vicissitudes, had guided them to the Promised Land. And, henceforward, they were thoroughly obsessed by history—seeing God as essentially the God of History—celebrating him in this aspect even more fervently than as the God of Nature. The whole course of history was envisaged as being based on a divine Promise, one which was conditional on fidelity to Yahweh, but was also a standing engagement, renewable indeed even after a breach of vows had had to be punished very severely, but capable of emerging this time even more impressive than before. All this helps to explain why so many of even the religious ideas of the Old Testament are so historical in character, and why the Bible is, to such a considerable degree, a history book.

When first the Roman Empire and then the whole continent of Europe went over to Christianity, their population inherited the Hebrew Scriptures as part of their faith, and came under the remarkable influence of the ancient Hebrew historiography. The fact that, in the meantime, Jesus Christ had emerged as a historical figure, walking familiarly in a world of ordinary things, had on the one hand tied religion more definitely still into mundane history, but also produced a narrative that was itself part of the constitution of the faith.

For fifteen hundred years and more, Western civilisation, even from its primitive stages, developed under the presiding influence of what we call an "historical religion." One of the results of this has been that Christianity has had its effect on Western views of history itself, even of general secular history. In the somewhat de-christianised world of the present day there has existed on a considerable scale, even in Europe, the view that men can make what they like of the past, fabricating such a picture of the course of things as may serve a present-day political end—one's narrative of the past being, in this case, avowedly dependant on what one means to do with the future. The Christian is particularly committed to the belief that events have a genuine significance and that it is necessary to learn the truth about them even when the truth is politically inconvenient, even when the truth has no purpose to serve. Truth itself becomes not a means to an end, but its own supreme purpose, therefore, the recovery of the past being a thing to be valued for its own sake, independent of all the utilities. The early development of history often seems to have been connected with primitive religion, and the passion for Truth as such marks the urgency of early man's desire to understand the meaning of life. (For this very reason, it is also true that religious belief at later stages in the story has often proved a powerful obstruction to historians.) But the notion of a Truth that is absolute, and the passionate need to press towards this Truth for its own sake is connected with an original urge to master the question of human destiny.

It might be a question whether historians in general or Christians in general do justice to the interest and the importance of the main themes of what we call Old Testament history. The literature at this point throws a remarkable light on our own cultural ancestry and on the basis of three of the world's great civilisations—the Jewish, the Christian, and the Islamic. It tells us of the origin and vicissitude of a nation possessing a disproportionate influence on the character of the future "Western Civilisation." Students of history require this story in its most scholarly form, the Old Testament account being subjected to a most methodical criticism and supplemented by all the other kinds of information that can be collected. Still more important, however, and more likely to appear relevant if it could be made communicable to more people, is the Old Testament narrative of what we might call our own religion through centuries of history on its own ground.

It might be held that, given the general character of the modern Western outlook, a short account of the rise of religion in the world, even the emergence of its primitive forms, but, most of all, the whole development of the spiritual side of man, down to the days when the ancient Jews brought some of their literature to the very edge of Christianity—all this is an important aspect of universal history not to be despised even by those people who question the reality of the spiritual, but who can hardly fail to testify to its power in human beings, hardly fail to want to show how, if they were illusions, they had come to prevail in the world. All this story and the whole tremendous quest into the spiritual which showed itself over a good deal of Asia, particularly perhaps in the centuries before the birth of Christ, might well be a greater support to the Christian faith in these days than, for example, too heavy an insistence upon the political history of the ancient Hebrew kings. It is a story to be told by the professional historian who collates the underlying religious sources with all the other kinds of evidence that may prove to be relevant, but will not close his eyes to actual manifestations of spiritual power even if he judges that power to be the effect of delusion—indeed, he will at least seek the explanation of the hoax. And the world has perhaps not sufficiently recognised the interest and importance of the fact that so much of the actual documentation, so much, for example, of the devotional writing, has survived, partly in the Old Testament, but not merely in our canonical Scriptures.

III

In a certain sense the seventeenth century, by a narrowing of vision and a concentration of attention, put an end to a lot of floundering, and achieved impressive results of a particular kind. They showed that, within the realm of things that are actually observable, man can develop wonderful systems of knowledge—the kind of knowledge that seems fated forever to multiply our power over the material universe. Yet, from another point of view, the very people who take part in these operations may still in a sense be prisoners of their technique, losing contact with other enriching forms of thought, and adopting in advance a posture that closes their vision to anything of a spiritual order. Even Catholics and Protestants can be brought to agree at the present day about the concrete evils existing in the Church on the eve of the Reformation. But, while the Protestant may argue that evils of this order justified the rebellious conduct of Martin Luther, the Catholic may reply that no abuses of this kind could ever justify breaking the unity of the Church of Christ. And this difference of opinion—momentous though it may be—is not strictly a historical one, not a matter to be settled by historical evidence or by mere technique. In a sense it pierces too much into the interior of things.

It is not possible for any of us to remain underground all the time, burrowing in a specialised realm of sheer technique. It is necessary for all of us to come up into the free air sometimes and to face the four winds as we make the sovereign decisions that require all our experience, all our sentiments, and the totality of our knowledge. For the most important things, I fix my eyes (not merely as a technical historian, but as a living and experiencing personality) upon the whole drama that takes place under the sun, having in mind its inner as well as its outer aspect, the experiencing self as well as the thing experienced. I am not sure that most people have meditated sufficiently on the relativity of our schemes of knowledge, and the way men's minds can come to be shaped to the techniques they habitually use. The smaller people, the ones most locked in their particular techniques, emerge as the most intellectually arrogant, precisely because, being unconscious of the limitations, they plunge forward without ever a misgiving. In reality all the sciences—certainly not excluding history—are prone to empire-building, prone to dogmatisms that are *ultra vires*. It appears to be impossible for men in any period of history to imagine that the constitution of prevailing thought in some future period is going to be different—different perhaps beyond all recognition—from that which is current in their own time.

Indeed, when we make the attempt to consider life in its wider aspects, and to fix those ideas by which we shall measure every-

thing else, we have to make a much wider call upon our experience—a more penetrating movement into our own interior, for
one thing—and we come to something very different from established historical knowledge, partly because it embraces both the
outer and the inner self.

It has always been my view that the pious men throughout the
Christian ages have not been cheated, and that the profoundest
insights do in fact come to those who, however much baggage
they may carry in their minds, can unload themselves of it for a
period, achieving an uninterrupted view, and becoming as little
children who walk humbly before God, not pretending that they
reign with their intellects over the whole cosmos. And those were
always right who felt this was not a dreary asphalt universe but
was warm with a Presence, who felt that God was intimately
with them, closer than their own shadow, tugging upon them all
the time with his love, and treating each of them as a special
case. They were even right when, if misfortune came, they wondered how much of this might be a chastisement, how much perhaps rather a process of refining such as is used for precious
metals, how much just a jolt given to them to open their eyes to
their mission in the world—God not being responsible for the
crimes of men, yet not a stranger to the tragedies that occur, not
contracting out or denying his part in the way the chips actually
fall. They, as Christians, do not expect less suffering than others,
but open themselves to more, not thinking themselves exempt
from the greatest tragedies, only sure that, through all of these,
they are brought nearer to Christ.

But what is forbidden is that this internal view—so intimate that
it is almost impious to throw it into the babble and debate of the
world—should be externalised into a crude mythical picture. We
may not, while pretending to be wearing something like the
thinking-cap of the historian or the scientist, start imagining the
spectacle of a God who stands above the stage of the world, pulling the strings in a kind of puppet-show, acting exceptionally no

doubt but really making himself only another factor in our mundane scheme of cause and effect—indeed, one might say, tinkering with the world's happenings to make them conform to a pre-arranged blueprint. Certainly one must never think of God as anything *less* than a man, but when we think of him as too like a *mere* man we are in danger of finding ourselves trapped into too many anthropomorphisms. If those who think of him as creating human beings for the achievement of some utilitarian object would only say rather that men were created for the glory of God, they would at least be speaking, and doing their thinking, in the right universe. Perhaps we ourselves comprehend God most adequately when not only outsiders but even we imagine that we have abandoned the business of trying to comprehend him—this, for example, when we have been seized by a great love of him, or have felt his warmth, or have achieved a state of unaccustomed restfulness. And perhaps there are only two classes of people in the world: those who believe that God created the wonderful universe that we know and those who prefer to believe that it is the universe itself which, in the multiplicity of its workings, is really creating God.

The Christian may too easily assume that he knows the mind of God, knows that one thing rather than another in the past is the work of Providence. And it is not to be expected that he will discover God's purposes for the world simply by brooding over longer and longer stretches of mundane history or putting given events under a bigger microscope. Perhaps, those who see primeval matter being shaped for millions of years into a habitation for living creatures, and these latter, after countless further ages, blossoming into the successive grades of intellectual life—culminating in a Newton or a Beethoven—will feel that, behind this story, there must have been the creative mind of a God who had had a previous plan of the whole course of things. It must be true at the same time, I think, that even the created universe itself has a dynamic of its own.

What was probably questionable in some of our older historical writing—especially in days when the knowledge of the way in which consequences proceed out of causes was still at an elementary stage—was the way in which the internal and the external view of events were allowed to become entangled with one another. There was a too facile assumption that one knew the mind of God, knew that one thing rather than another in history was the work of Providence. Against this, one might say that every student of the actual past must carry the scientific method—or the pseudo-scientific way of doing history—just as far as it can go, doing this, however, without arrogance of mind and recognising that here on earth we are only looking at things from the underside. A man may feel that in the intimacy of his own inner life he has seen the work of Providence, or felt its inescapable pressures. This belongs to "internal" knowledge or experience, and it may lie as a secret between the man himself and his Maker, hardly to be regarded as ordinarily communicable to the outer world, although an authentic thing, and recognisable to a fellow-Christian, especially to one who is well-endowed with empathy.

Historians have turned sometimes to the question of Providence, for they are well aware that they have never exhausted the possible lines of causation and, even on an over-all view, one never quite knows where the chips are going to fall. Some have worked out the shape of the leading movements of history—the rise and fall of empires, the wars and revolutions, the renaissances and declines—and have seen these as themselves the over-all map of the dispositions of a Providence that is distributing praise and blame. There are some who, finding a gap in the fabric of historical explanation (or of scientific explanation) see this as a kind of loophole for bringing God into the question. This is a thing that one often notices (without necessarily quarrelling with it) in the work of the famous German historian Ranke. Schlegel took the line that the historian ought to be very shy and reticent, giving a mere shadow of a suggestion to support the idea of the

possible working of Providence in a special case. There are some
who bring in Providence where the ancient classical writers might
have the conception of "chance"; and perhaps for those who are
believers in God this is an understandable transmutation though
hardly an adequate policy. So far as concerns the practical world
I think I am sympathetic with those people who say that, if you
are engaged in some transaction you should both trust God and
keep your powder dry. Do your best, but accept the fact that
the crucial result may be in other hands. In other words, believe
in Providence, but remember that Providence may be calculating
on working through you—you always have to do your best in
any case, but working with Providence you may be less hampered
by anxiety about the result. We ought not to be frightened into
believing that it is meaningless to thank God for the coming of
Spring or for bringing us through another day. But, even about
this, one need not be too rhetorical. Perhaps it should be rather
like holding one's breath for a moment, holding one's breath before the beauty of the thing.

IV

When I try to consider the question how far being a Christian
should affect people in their writing and their thinking about
history, I come to the conclusion that some of the better principles of our historical writing, though they have been secularised
now, are really the result of a civilisation soaked for a thousand
years in basic Christian teaching. I think the view that the men of
the past—the dead—have one claim upon us, the claim to be
understood, not just to be blotted out as fools or rogues or vested
interests because they did not agree with us, would be an important example of this. I mean also the view that men should be
assessed in terms of the outlook of the age in which they lived,
and not condemned because in the sixteenth century they failed
to understand our twentieth-century idea of democracy. It is a

very difficult thing for men to rise above the prevailing ideas of their age—indeed at a given time only a few people can be as original as that. The greatest men can be original only about a few things, and they follow the prevailing ideas of their world in all other respects.

I think that a Christian historian ought to give something of himself in order to reach a profounder understanding of historical personages, accepting the fact that a colossal effort of understanding is required, and that all are sinners and, in reality, judgment belongs only to God. If the historian is a Christian, one might expect him to take his role with some humility, and not to imagine that he is a sovereign who pretends to possess the last word, or to decide the judgment of posterity, but remembering that he is part of the very drama he is dealing with, often also a partisan in any conflict. He is at his best when he sets out first and foremost to explain, to account for, the way men behaved or came to be what they were. And, though while Hitler was alive it was necessary to fight him and stop his cruelties and destroy his power, still it would be good if an historian could explain to us how a boy of ten, playing in a city street, could have come to be like that—the historian behaving like a mother who sadly has to observe how a son of hers is going wrong.

All the same, the Christian historian would have to take the line that, however much explaining one does, human beings do in fact have some real responsibility for what they do and for the kind of persons they become, also the kind of world they are helping to create. And we write our history with this in mind, since men are not mere puppets, not mechanical dolls—but each is a separate fountain of life, an independent source of action, a starting-point of original things. Even the history of the French Revolution cannot properly be recovered or understood unless we see individuals who may *will* something but might well have willed something different—individuals in action, producing their interplay, and altering the very course of things by their loyalties, their angers, their cross-purposes and their intrigues. In other words, even those who are studying at the ordinary mundane level must have a humanist kind of history, the sort that is a narrative of human beings, not merely a study of forces and social movements and generalisations. History, indeed, is a thing which can never properly be reduced to mere process. There is something in history to which we fail to do justice unless we present it as a story in which men never quite know what is going to happen next.

9

The Christian and Historical Study

I

Some people seem to deplore what they regard as the anarchy of ideas and systems which exists in the world at large, and even within a single university at the present day. They regret the lack of that uniformity of Christian outlook which characterised the Middle Ages; or they hanker after the kind of unanimities that are supposed to underlie a Communist order. Romanticising about the harmony of a regime in which there exists this happy agreement about fundamentals, they close their eyes, however, to the ugliness of the methods by which such a system has to be achieved. They overlook the fact that perhaps only at certain periods in the history of a civilisation can such a thing be regarded as achievable at all.

One important fact seems to condition the history of modern thought—a fact that is rooted in the very constitution of the universe. It is the fact that unanimity is easy to reach amongst human beings on concrete questions very near to earth, such as the question of the effect of heat on the substances to which it may be applied; but as soon as one moves to a higher level, to the question of the nature of heat itself, for example, or, higher still, to the question of the origin of all heat, human beings seem to differ more widely the more lofty the intellectual region to which the matter is carried. Where the mind of the group predominates and the spirit of the herd keeps men close together a certain solidarity can certainly be achieved for a time even in the higher reaches of thought or speculation. But if it is not true that civilisation always moves towards a higher differentiation of personality—and towards a greater respect for individual people, or for the options they make at the highest levels of thought and decision—it is certain that a Christian civilisation must move in this direction, especially as the element of voluntariness is bound to be an important factor in a religion so personal and intimate.

A Christian civilisation, precisely because it must embrace so high a conception of personality, must move towards what Christians themselves may regard as its own undoing—towards freedom of conscience instead of greater solidarity in the faith. A world in which personality and conscience are respected, so that men may choose the god they will worship and the moral end they will serve—this, and only this, is a Christian civilisation when human development has reached a certain point. Precisely as such a civilisation becomes more advanced, the problem of securing uniformity in the most lofty realms (such as that of religion) becomes more overwhelmingly difficult. It is a fundamental part of the case against Communism that it puts back the clock in this respect (or perhaps that it flourishes only where humanity is in a backward state); and in any case, since it cannot be argued that unregenerate man is naturally Christian, it is bad tactics as well as bad ethics for Christians to dwell too greatly on the advantages of uniformity as such. Uniformity would only be too likely to come at their expense—the unbeliever treating them to those severities which, so long as they had the power to do so, they meted out to him. It is legitimate, then, for Christians to hope to convert as many people as possible, but not to translate this hope into a dream of terrestrial power, or to expect from Providence that the dice shall be loaded in their favour and the forces of the world itself ranged on their side. We must first praise God for the human intellect and the freedom of the mind, and only after that is it legitimate to pray that men—as free men—may come into some degree of unison. Christians are strongest if, regarding themselves as the servants rather than the masters of men, they claim no peculiar rights against society as such. They must claim the right to worship the God in Whom they believe, and they have no justification for regretting that others should have the same freedom in the matters that most highly concern human beings. This tolerance is the minimum that we must have to make a Christian civilisation.

This being in fact the situation, whether we like it or not, we do not despise the liberty that so many generations fought for; nor shall we risk adding to the dreadful mistakes which ecclesiastical authority made in the past in respect of this matter. We do not deplore the anarchy and variety of ideas and systems, but ask only that the sparks shall be kept flying—ask indeed that there shall be more scepticism in both Christians and non-Christians over much of the spacious realm of thoughts—hoping that, if the non-Christian is not converted, even he may be induced to greater elasticity of mind. In particular our educational institutions ought not to be quiet pools, where intellects shall comfortably settle down, but a seething cauldron of ideas, a fair arena for the clash and collision of intellectual systems.

It is in this kind of a world—a world where Christianity as in New Testament times stands on its merits against a more or less neutral or pagan background—that we must consider the attitude of the Christian to history as a branch of scholarship. We will take account of the whole question of the scientific method as such, and also of the peculiar features of history as a human study. Then it will be relevant to consider the question of the possibility of a Christian interpretation of the past.

II

Over a considerable part of its area history may be conceived to be a science in the sense that it studies very concrete and tangible things, such as can be tested and attested by a definite kind of evidence. Furthermore, it examines the observable or demonstrable connections between those things—the relationships between various kinds of what we call historical "events," for example. Technical history, on this definition, is to be regarded as a mundane and a matter-of-fact affair, and serves only limited purposes. It may provide us with a demonstration that Jesus Christ did live or that a certain saint died at the age of fifty; and if it does prove those points it proves them for all men, whatever their faith—its argument is valid for Catholic or atheist, for Marxist or Mohammedan. There are many things, however—and those much the most important—which the technical historian knows that his evidence and his apparatus give him no special right to decide. Amongst them we should include the quality of Beethoven's music, the rightness of the Reformation and the question of the divinity of Christ. When the technical historian explains the victory of Christianity in the ancient Roman Empire, we should not expect him to say that the success was due to a decree of Providence or to the authenticity of the religion itself. We should rather expect him to provide an empirical study of certain tangible things that gave Christianity its efficacy in the world of that time. There would be many cases where the historian would be aware that he had not found the clinching argument, or fully established even so concrete a thing as a date, or demonstrated his hypothesis to the satisfaction of all his fellow-students. This kind of history, therefore, ought not to appear as a self-complete intellectual system or as a continuous piece of explanation without any holes in it. In reality it is merely the extension of the universal habit of men to reflect on the observable connections between events—beginning, one might say, with the daily rising of the sun, or the experience of the trouble that is likely to be provoked if one steals one's neighbour's food.

If it is asked how this initial view of history—this view of it as a science—is connected with Christianity, or it is argued that so mundane a conception of the subject is actually inconsistent with religion, one may reply that on the contrary there are reasons for suggesting that this approach to any science is a specifically Christian one. It is the view which comes from regarding the historian as a person under a certain kind of discipline for the purpose of examining the ways of Providence and the structure of the providential order. It does not deny Providence. It does not hold that events will form a self-explanatory system without any necessity for the idea of God. It relegates scientific history to a humble rôle, therefore—certainly not assuming that the study of demonstrable events will suffice either to answer the question whether the hand of God can be found in history, or to explain why man exists, or to settle ultimate philosophical problems. And certainly it does not assume, as the Marxists and so many other secularist thinkers seem to do, that when we have learned the history of a thing we shall have achieved its final and total explanation.

The scientific method that we are discussing seems on the whole to have come into existence in the way that has been described—both the natural scientists and the historians acting in the belief that they had found a better means for studying the ways of Providence. And this affected the situation in a definite manner, for it meant that they felt themselves to be operating only on the outer fringe of something far bigger than their instruments and observations could reach. They felt a greater distance between the kind of thinking which analyses a blade of grass more and more minutely or observes the stars over wider and wider expanses of space, and the kind of thinking which estimates the nature of the universe or judges the meaning of life or decides the question of the existence of God. It was discovered that by restricting oneself to the realm of secondary causes, one could pursue certain kinds of more mundane enquiry to better purpose. This is at any rate one of the secrets of the transition to the scientific method of modern times.

It appears to be the case that the scientific method gained great impetus from the fact that students of the physical universe recognised the search for final causes to be too difficult as well as too distracting for the particular purposes of physical enquiry. Here, as in history, the restriction of scientific attention to the analysis of tangible things or their inter-relations freed the mind for a more specialised form of research and released the thing we now call science from its entanglement with all the pretensions of a "natural philosophy." The case for the scientific method was strengthened by the discovery that students of differing philosophies and religions might discuss Nature to better effect if they shelved ultimate questions and debated more tangible things— things which, when once established, were established for men of all persuasions, so that there was common ground for intellectual interchange. At the next stage in the story, and almost sometimes apparently in absence of mind, men came to imagine that final causes had been disposed of, and Providence eliminated altogether; and this meant that one had locked oneself inside the scientific method, so to speak, forgetting the terms on which one had got into it in the first place. It was a modern piece of wilfulness which made men think that technical history and natural science were qualified to settle ultimate philosophical questions for us while they themselves were in an interim stage, as they still are, nobody knowing what surprises they may bring at the next turn in the road. When Sir Isaac Newton established the laws which govern the relationships between the movements of the heavenly bodies, some men slipped too easily into the view that the entire universe was a "mechanical" affair—even chemistry and biology became too mechanistic. Like the mathematician who discovered that God worked as the Great Mathematician, they ran too quickly from the conclusions reached in a certain field of science to over-all assertions concerning the whole order of created things. In reality the very factor which gave the scientific method the advantage in efficacy and intensity and *élan* was the restriction here described—the restriction of the scope of physical enquiry itself.

If the historian becomes interested in processes and laws or finds subtle conformities of pattern in the movements of events, this, too, is not to be regarded as uncongenial to the Christian outlook. Regularities in the universe were discovered very early, and even in Old Testament times it would have been no marvel for God to stop the sun if the normal motion of the sun relative to the earth had not been an accepted idea. It is possible that the belief in the existence of a coherent world-system was encouraged by monotheism and that the search for rationality in the universe was furthered by the fact that men believed in an intelligent Creator. Some seventeenth-century scientists seem to have felt that Creation itself would be imperfect and the rationality of God in doubt if the physical universe could not be envisaged as a perfectly dovetailed system of laws. Even in the Middle Ages men were aspiring to discover the very kind of laws which Sir Isaac Newton ultimately put forward for explaining the heavens and elucidating the problem of motion. It almost appears that religious minds could hanker after laws and rationality even before the modern scientist had found a more adequate way by which to discover the form of the laws themselves.

I do not quite know what this realm of law is, under which we see the universe operating, or how far it represents merely some kind of rapprochement between our limited reasoning and an external world which we only partly envisage at best. Sometimes I think it is like the case of the man who wanted to cut up a piece of soap for analysis, and, having used a square-shaped potato-cutter for the purpose, ended by discovering that squareness or squarity or the capacity for being cut into squares was the essential quality of soap. For those who have forgotten the origin and the terms of the method I have been describing, technical science and technical history may become like tightly woven screens, all the threads of which are interlocking, all the meshes drawn as close as possible, to cut the student off from any outer light. And we must not exult too readily if there seem to be some holes in the screen—we must not say that this is something unaccounted for, so it must be God—for the scientists if not the historians may answer that with the acquisition of further knowledge they count on closing up that particular kind of hole. It is possible to hold that scientific explanation, though a limited thing, can conceivably be an unbroken fabric so far as it goes—in other words, can, within the limits, be self-complete.

We might imagine ourselves locked in the system that natural science and history fasten around us if there were not one glaring hole in the screen—a hole which nobody can ever pretend to patch up. There is something which is closer to us, more intimate, more real, more direct, than all the external evidence in the world. The only thing in the universe which any of us can know in any sense from the inside is a single personality—namely, himself; and only from an internal knowledge of ourselves can we begin to build up our impressions of other people. The primary judgment that any of us makes, anterior to all philosophising and all scientific endeavour, is a judgment that conditions all other judgments—namely a judgment that we make about ourselves. The historian, in this particular sense, does not regard personality as a mere "thing," to be studied as other external things are studied. At this point, as we shall see, he rises above what are generally regarded as the ordinary methods of science.

It may be suggested that, though religious men have been inconceivably unwise on so many occasions, the Christian who adopts the view of the scientific method that has been described is in a position to hold his mind more free for hypothesis than those who seek from science their over-all view of life and the universe. It is the Marxists and the secularist systematisers of our time who, without reaching as high as God and without confining themselves to necessary inferences from observed phenomena, commit their minds to vast intermediate systems of ideas—systems which are less capable of elasticity than science itself demands, and which control the range of hypothesis sometimes, or constrict the adventures of the mind, since they create their own demand for conservatism and consistency. On one occasion in the Middle Ages, when the teaching of Aristotle insisted that God himself could not have created a vacuum or an infinite universe or a plurality of worlds, an ecclesiastical decision rejected this limitation upon the power of God, and freed these things for hypothesis in a way that the Church has too rarely dared to do. The believer in Providence can be prepared for any surprises. The Christian need put no limits to the Creator's versatility.

One cannot even feel certain that the view of truth to which we are accustomed would long survive the existence of a Christian civilisation. It may still transpire that the notion of what we might call absolute truth is not unconnected with religion—not unconnected with the feeling that here is something which at any rate you cannot cheat God about. One is tempted to suggest that there has been perceptible in Marxist propaganda a view of historical (and even scientific) truth which, if it were developed over a long period, would have an unfortunate effect upon the pursuit of knowledge and the very conception of scholarship. It is difficult to be sure what safeguard there would be against such a "utilitarian" handling of truth if the world were to go on becoming increasingly pagan or at any rate increasingly materialistic in its preoccupations and ends. The ideals of "academic" history may transpire to have been the legacy of a Christian civilisation after all. If Christians themselves when writing history have sometimes been too intent on a kind of "truth" which attracted them merely because it seemed to serve a good purpose, they have operated, as Christians no doubt often do, against the tendency and the ideals of Christianity itself.

On the whole issue of the scientific method we may say, then, that if it is not a specifically Christian thing, it did at any rate develop in the heart of a Christian civilisation. Many of the people who actually developed it maintained that they were glorifying God in the very pursuit of their researches. The case was different from that of Aristotle and the ancient Greeks precisely because the science had specialised itself out, and was no longer rolled up with natural philosophy. It is not clear why there should be any conflict between religion and the scientific method, unless the religious man is too materialistic or the scientist becomes arrogant. And modern science has been beneficial for Christianity in that it has forced Christians to disentangle their faith from theories of the material universe; it has made religion cease to be plausible except as an essentially spiritual thing. On the other hand, the essential effect of Christianity on the form of scholarship that has been described is one not merely congenial to science itself but absolutely indispensable to it—namely, to increase the necessity for intellectual humility. The worldliness of modern man is not the result of the devising of the modern scientific method and is possibly rather a characteristic of modern urban civilisation. And if we say that at least the scientific attitude gives great leverage to the worldliness of worldly-minded men, it can equally be said that it gives great leverage to the spirituality of spiritual men, for it is in itself a neutral instrument. Even if they are not scientists or historians, men do occupy much of their day in some sort of commerce with concrete things. The director of a cathedral choir who forgot the importance of worship, and lived rather for music as such, is seduced by the world in the same way as the student who makes science the very end of all his endeavours and the sum of all his interests.

III

If history is of some importance by reason of the things which it can establish scientifically, however, large areas of it are by no means so positive as this phrase would suggest to many people. In a similar way we might say that large areas of it are less capable of reduction to regularity or law than many people would seem to wish. History tends to differ in its whole organisation from anthropology because it gathers itself to such a degree into stories about personalities. And Marxist history, which sometimes seems to aspire to something like the organisation of anthropology, surprises and shocks many people because it looks less like life—it is so much a matter of process and schematisation, so little interested in the individual.

The kind of history which has developed in our civilisation and was handed down to the twentieth century has clustered around personalities and we have tended to think of it as organising itself into the form of narrative. It resurrects particular periods, reconstitutes particular episodes, follows the fortunes and discusses the decisions of individual people, and rejoices to recover the past in its concreteness and particularity. It does not limit its interests to the things that can be reduced to law and necessity—a project more feasible to those who direct their studies upon the materialistic side of human beings and human purpose. It is more interested in what is free, varied and unpredictable in the actions of individuals; and the higher realms of human activity—the art and the spiritual life of men—are not inessentials, not a mere fringe to the story. The play of personality itself is not a mere ornament in any case—not a kind of cadenza or violin obbligato—but is itself a factor in the fundamental structure of history. The historical process is so flexible that all the future would have been different in a way that it is beyond the power of our

mathematics to calculate if Napoleon had been shot in his youth
or Hitler had failed in January, 1933. In this sense history is like
life and every individual should be aware that it does really matter to the world what decision he makes on a given issue here
and now.

Now, this attitude to the study of the past, if it is not to some
degree the effect of our traditions—our Christian civilisation,
with its high view of personality—is particularly congenial to
those traditions and particularly appropriate for the Christian. It
implies a telling of the story which has the effect of doing justice
to freedom as well as necessity, and in which the spiritual (as well
as the material) is organic to the theme—not a mere added ornament. It is typified in the flexibility of narrative, and is to be con-
trasted with the kind of history which sets out rather to schema-
tise the centuries or turn everything into a process. The traditional
historian has shown an interest in individuals for their own sake,
and in a bygone generation as an end in itself, which we in our
civilisation have perhaps too easily taken for granted. It is possible that a grossly materialistic civilisation, too intent upon utilitarian purposes, would not see the point of these things and would
not produce the kind of fabric that we call history. The Christian
must defend it however; for this is a kind of history in which—in
a certain sense at least—personalities are the irreducible things.

Our traditional historical writing has gone further than this. It
has refused to be satisfied with any merely casual or stand-offish
attitude towards the personalities of the past. It does not treat
them as mere things, or just measure such features of them as the
scientist might measure; and it does not content itself with merely
reporting about them in the way an external observer would do.
It insists that the story cannot be told correctly unless we see the
personalities from the inside, feeling with them as an actor might
feel the part he is playing—thinking their thoughts over again
and sitting in the position not of the observer but the doer of the
action. If it is argued that this is impossible—as indeed it is—not

merely does it still remain the thing to aspire to, but in any case the historian must put himself in the place of the historical personage, must feel his predicament, must think as though he were that man. Without this art not only is it impossible to tell the story correctly but it is impossible to interpret the very documents on which the reconstruction depends. Traditional historical writing emphasises the importance of sympathetic imagination for the purpose of getting inside human beings. We may even say that this is part of the science of history for it produces communicable results—the insight of one historian may be ratified by scholars in general, who then give currency to the interpretation that is produced. A Thomas Carlyle might convince us that he had found a clue to Cromwell, and yet he might fail to carry us with him in his reconstruction of another person like Mirabeau. The whole process of emptying oneself in order to catch the outlook and feelings of men not like-minded with oneself is an activity which ought to commend itself to the Christian. In this sense the whole range of history is a boundless field for the constant exercise of Christian charity.

IV

At this point it becomes relevant to discuss the possibility of a Christian interpretation of history within the scheme of things which is now in question.

What we begin with is a form of historical scholarship restricted to a realm of tangible things, things which are to be established from concrete kinds of evidence. It is necessary to make inferences from the evidence and to have insights into personality, but the inferences and the insights belong to the same limited realm; they are, so to speak, very near to earth. In fact, we should expect them to be generally communicable, indeed to be ratified by a certain consensus of opinion, before the result could be accepted as a part of scholarship. In all this we may feel that we are studying the ways of Providence, but we cannot say that we have demonstrated the existence of Providence—we cannot say: "Here is evidence that ought to be sufficient to convince any neutral person." When we have reconstructed the past all that we have obtained is a picture of life as it must appear to any person living in the world; except that, whereas an individual only sees his three-score years and ten of it, he can now extend his vision and recognise certain long-term processes and tendencies.

If in life a man has accepted the Christian view of things, he will run these values throughout the whole story of the past, and, taking the very basis of narrative which historical scholarship has provided, he may see every event with an added dimension. He will have used historical science in order to become a closer and better student of the ways of Providence. He will see the vividness and appropiateness of the Biblical interpretation of history for the study of any country in any age of its history. He will not claim that historical science has demonstrated the truth of the interpretation which as a Christian he puts upon human events. His over-all view of things is partly dependent on the attitude that he brings to history in the first place; and partly it is dependent on the most intimate judgments that he makes about himself, about life as he has experienced it, and about the course of centuries as he has gathered it from historical scholarship. In this sense there is a Christian interpretation of the whole human drama, which is simply an interpretation of life—indeed, an aspect of the religion itself.

It is often assumed, however, that within the field of historical scholarship as we have described it, there is a Christian organisation which can be given to the narrative; in other words, that the European history which appears in our educational curriculum can be given such a form that it bears a Christian interpretation and vindicates the Church not merely in its spiritual functions but in its mundane policies. It soon becomes apparent that there has to be a Protestant history which is not only different from the Catholic version, but violently contradictory to it at times. Indeed, the attempt to vindicate Christianity in history easily turns into an attempt to justify Christians instead.

In the particular realm with which we are concerned at this point in the argument it is possibly unfortunate if the student of history ever ceases to regard himself as essentially an enquirer. Nothing so hardens the mind as an attachment to a particular reconstruction of history which we may have motive for cherishing because it establishes a case. It is true that the mere yearning for historical comprehension has sometimes seemed to be an insufficient motor for the mind, and has left scholarship lacking in one way or another until (as Lord Acton once pointed out) sheer polemical passion has brought a relevant fact to men's attention or has driven a violent partisan to an original piece of exposition. A Christian writer who possessed such polemical fervour might call attention to something which previous historians had overlooked. If he established his case the matter would become a constituent of historical scholarship in general, however; it would no longer be the mark of a specifically Christian view of the past.

It would perhaps be regarded as legitimate to envisage the history of Europe as the story of a civilisation which developed under the presidency of the Church and which for many centuries bore an unmistakably Christian stamp. It would be necessary to be very cautious, however, in any attempt to make polemical use of this particular formulation of the narrative. In other societies and other regions there have been other religions which have presided over the development of a civilisation; and sometimes the parallels, as in the case of Islam, are very remarkable. If we may infer that at certain stages in the development of a civilisation religion can have a presiding rôle in the story, we can hardly turn this into an argument for ecclesiastical predominance in entirely different stages in the history of human and social development. Furthermore, in the form in which we are now considering the case, it is difficult to see how it can be used as an argument for one particular creed as against another.

It is possible to produce a form of history which gives the Church the credit for all the good that is done, but, when wrong decisions have been taken—however disastrous for a generation—attributes these merely to the defects of the human agencies, so that "the Church" is always right, whatever men may do. On this view, every ecclesiastical decision ever taken about mundane affairs might be wrong—might even be recognised to be wrong—and yet "the Church" itself never come within the range of criticism. Such a form of argument would have its dangers if in various respects and in the affairs of the world it were difficult for people to hold fast to the distinction between "the Church" and its human agencies. Ecclesiastical systems, in the form in which they confront the historian, have their aspect as very tangible, astonishingly human, systems. An optical illusion concerning them would serve the purposes of real live men who might be desiring an illegitimate form of power.

We are saved from some of the optical illusions if we say that Christianity itself is always right and it is just the Christians who, because they are only human, tend to go wrong in history. Yet an hostile observer might argue that an exclusive religion can even produce terrible evils in the world unless accompanied by a great amount of charity—a greater amount than has sometimes seemed to be available. He might argue that, where charity is defective, there is liable to be a special rigidity in religious morality or convention, a rigidity which has made Christians or ecclesiastics lag behind other people sometimes on ethical points on which churchmen of the twentieth century have come to lay great store. An hostile observer might even point out how the insistence upon what is due to God—in the matter of actual property and wealth, for example—has been used to cover ecclesiastical interests of an extremely earthy nature, and even abuses of a glaring kind.

Three things, however, seem to illustrate the importance of Christianity in that mundane history which is under discussion—the importance of the particular religion which presided over the rise of what we call our Western civilisation. They all spring from the very nature of the Christian gospel itself and their effects on our civilisation are merely the incidental results of the ordinary religious activity of the Church—they are not a sample or a vindication of the mundane policies of ecclesiastics. They are by-products of the missionary and spiritual work of the Church, and it is not clear that the same mundane benefits would accrue if men set out with the object of procuring the mundane benefits—if men worked with their eyes on the by-products themselves. They show that the Church has best served civilisation not on the occasions when it had civilisation as its conscious object, but when it was most intent on the salvation of souls and most content to leave the rest to Providence. The three things are the leavening effect of Christian charity, the assertion of the autonomy of spiritual principle, and the insistence on the spiritual character of personality. Apart from the softening effect that religion often (but perhaps not always) has had on manners and morals, these things have had their influences on the very texture of our Western civilisation.

In this sense Christianity has operated through the centuries to greater effect than Christians themselves have done, to greater effect indeed than Christians sometimes seem to like it to do. It has operated for liberty even when Christians are opposing liberty and even when ecclesiastical authority has been the very enemy that has had to be fought. It is remarkable to see how greatly modern freedom has risen out of Christian history and modern liberalism out of medieval politico-ecclesiastical controversies. It is remarkable to see how many of our freedoms have been built upon an initial religious claim for liberty. In a similar way, the assertion of the autonomy of the spiritual principle in the Middle Ages prevented anything like absolutism and produced in society a friction that was stimulating. The conflict between ecclesiastical and lay authorities allowed in any case greater play for the individual than a completely totalitarian system admits of. In modern times a Christianity in opposition—the Nonconformists in England for example—inherited the claim that had been asserted on behalf of the spiritual principle, the claim that it should stand on its own feet and if necessary not merely oppose the government but criticise the very form of society on moral grounds. This Christianity in opposition is the father of many of the modern movements of social and political reform. It might be objected that many of the liberties and reforms which we in England particularly prize, and which the West claims to represent as against the Soviet system, were largely developed, and largely made effective, only after the period when our Christian civilisation had become predominantly secular. Even the reformers who came into conflict with the Church after 1700, however, did not realise to what a degree they had merely secularised many features of the Christian outlook, though they imagined that they had entirely cut loose from Christianity. As twentieth-century pagan civilisations develop their barbarities we shall realise more and more what—even amongst non-Christians—has been the leavening effect of Christian charity and the Christian outlook.

If there is anything which the Christian might peculiarly feel about our European story—though he could not scientifically prove it or expect his view to be shared by those who do not share his beliefs—it is an impression of the liberty and spontaneity and originality of the spiritual factor in history. For many of the enemies of Christianity this picture is hidden by the rigidities to which religion is liable when the spiritual factor is defective, or by the tendency of ecclesiastical systems to slide into routine. Yet non-Christian historians have done justice to the amazing power of spiritual men, though holding that these were defective in their self-examination and deluded in their spiritual interpretation of their inner life. It is important to note that it is the material world that is under the dominion of necessity. As States become more materialistic in their objects and purposes, State-policy itself becomes more powerfully dominated by necessity, and countries must cringe and cower under civil services becoming ever more unimaginative, more constricted by that environment which now controls men instead of being controlled by them. But men who have had Christian love in their hearts have been carried to original courses, driven to surprises and eccentricities. Those who have asked, "How shall we worship God?" have had a spirit that thrust itself out into Gothic cathedrals; they are liable to be at least more original and refreshing than those who say, "Why shouldn't we have some art, like the Greeks?" Where Christianity subtly affected the character of our civilisation—just here is the point where our world might have been different altogether supposing another religion had conquered the Roman Empire. The ultimate vindication of the Christian religion in history, however, is not to be found in any of its mundane by-products, but in the spiritual life itself. Because this is so intimate a matter we discern it better perhaps in biography than in European history as seen in the large.

V

It cannot be too greatly stressed that history is a thing which requires to be taught in totally different ways at different levels. The method most adequate to its purpose is perhaps the mere telling of stories to the very young, with possibly a side-glance at some moral that may be drawn from the narrative. One curious piece of inflexibility has possibly been ruinous to history as an education—the idea that there exists an ideal scheme or map of history and that between students aged nine, fifteen, twenty and twenty-three the only difference in the history to be taught is a difference in the scale at which the map is reproduced. When boys of nine are being taught the structure of the feudal system or sixth-form students are being pressed into work of university character in advance of their status, great opportunities are being missed, and the fallacy is like that of the ancient painters who presented the infant Jesus as a man merely drawn on a smaller scale. It is possible that for some young people history will be useful as containing examples from which a certain amount of political teaching may be elicited. Since history, however, is so intricate a network, with everything so entangled with everything else, it is not clear that sets of model instances, a modern equivalent of Æsop's fables, would not serve that particular purpose more adequately. The book of Isaiah may be used for devotional purposes or, alternatively, as an historical document, but it is not clear that the latter alternative will be the useful one for everybody. And it could hardly be claimed that every man should be trained in technical history, especially as even this training does not make a person really competent to range over all the centuries. Bury was a great scholar, but when he wrote on the nineteenth-century papacy he made howlers which would prevent an undergraduate from gaining the bottom class in a university examination on the subject.

Owing to the way in which the world has naturally developed, owing to the pressing necessity for so many kinds of technique, owing to the fact which we have already noticed—namely, that men can agree on the subject of the effect of heat on water while they come to differ radically when they rise to higher questions—owing to these things, the most important sides of education, and particularly the training in values, the communication of an adequate world-outlook, have come to be for the most part outside the framework of an ordinary curriculum. This is not entirely to be deplored, however, if the fact is faced and recognised, and especially when the family plays its due part in the education of the young; for we should hardly expect a formal education to teach the most important things, such as the aesthetics and the strategies of falling in love. It might be understood, however, that when Christians are teaching even the strictly technical kind of history, they in particular would remember the limits of the science, the need for humility of mind, the importance of getting inside human beings, the call for charity, the dynamic quality of the spiritual factor in history. It might be understood that they would see English history as part of the story of a Christian civilisation rather than a self-contained world, to be described with patriotic innuendoes.

It remains true throughout, however, that in teaching or reading or writing history the richest wisdom and the finest educational nourishment comes from the things which we generally describe as *obiter dicta*—the comments that are made in aside, the places where private views and the results of personal experience leak out, the things, shrewd and intimate, that a teacher throws in just for love. Once again we need not complain—we may actually rejoice—that the finest things in education must come from the creativity of the teacher himself and are extra-curricular by necessity.

The microscopic study of the transition to the "mechanistic" idea of the universe in the seventeenth century has shown to what a degree at the crucial moment men like Mersenne were moved by the argument that divine miracle could never be justified unless it could be shown that the world in its normal processes was regular. The truth was that one of the chief things that they had to fight was the current high-brow view of the universe as quasi-magical in character—a place where everything was so to speak a "miracle." Paradoxes like these are the parables of history and they illustrate the manifold ways in which religion itself helped to bring about modern sanity in respect of the purely material world. Let us be quite clear that in the field of history the Christian should be the first and the most extreme in demanding the scientific attitude; even though men may still differ so greatly in the evaluations which they place upon the results of the scientist's researches.

IO

The Christian and Academic History

The Case Against Academic History

It has often been argued in recent years that the modern study of academic history—and particularly the modern way of teaching the subject to undergraduates in universities—is unsatisfactory, because it provides no interpretation of the whole drama of human life in time, no explanation of the place of man on the earth, no philosophy or religion that will be of use in the world of actual experience. It has further been objected that if the student seeks, in the approved historical manner, to enter sympathetically into the mind of the twelfth century as well as the nineteenth, or into the outlook of Christian and Mohammedan, Protestant and Catholic, Whig and Tory alike, then he finds himself trapped into a kind of relativism, and he will not extract from historical study as such the kind of values which an educational system should induce him to acquire. There has occasionally been observable amongst Christians almost an envy of Marxist history—almost a desire for a kind of history which can be studied to some purpose, and which will itself give a man a purchase on events, a kind of history not controlled by a superstitious adherence to the idea of the past-as-it-actually-was, but governed rather by what one has determined to do with the future. Some religious thinkers appear to have gone further still, claiming that, since the Christian view of history culminates in the Cross and Resurrection, there is no purpose in studying the course of secular history, especially as it had added nothing to the real meaning of things in the last nineteen hundred years.

Here are four important objections which have their greatest force in respect of the teaching of history to the young, but which are often extended in fact to make a case against historical scholarship as such. They are arguments sometimes adduced against that kind of "academic" history which (instead of offering a total interpretation of human life and vicissitudes under the sun) merely seeks to lay out the story in its concreteness and its detail, and in fact only provides a limited realm of explanation, the evidence and the apparatus of the historian not qualifying him to claim a higher authority or function than this.

We may remove some obstructions, calculated to hinder or distract the discussion of the essential issue, if we admit that, granted the case for historical scholarship in general, there may be objections (serious for the non-Christian as well as for the Christian very often) to the curriculum and the mode of teaching which are generally in vogue in schools and universities. Further than this, many of us might agree that there is a Biblical interpretation of history which offers to the Christian the only satisfactory way of regarding the drama of the centuries, and discovering the attitude and role that human beings ought to adopt in relation to this. It is possibly true that the customary way of envisaging this human drama, the prevalent attitude to the whole course of history, is in these days an even more serious obstruction to the acceptance of Christianity than the current notions concerning the natural sciences. In any case if men have found no philosophy or religion in their actual experience of life it can hardly be claimed that the "academic" study of history will itself provide the remedy, or that the attempt to learn more scientifically when things happened and how they happened can solve the whole problem of human destiny or achieve anything more than a better statement of the essential riddle. When we have reconstructed mundane history it does not form a self-explanatory system, and our attitude to it, our whole relationship to the human drama, is a matter not of scholarship but of religion—it depends on the way in which we decide to set our personalities for the purpose of meeting the whole stream of events.

The Relation Between Historical Enquiry and the Interpretation of Life in Modern Society

For the assured Christian these problems—these objections to "academic" history—can hardly be said to exist; for, having in his religion the key to his conception of the whole human drama, he can safely embark on a detailed study of mundane events, if only as an examination of the ways of Providence. If "academic" history cannot provide a man with the ultimate valuations and interpretations of life under the sun, neither is it generally competent to take them away from the person who actually possesses them; and if there is internal friction and tension when the religious man puts on the historian's thinking-cap, the strain is just as constant between religion and one's actual experience of life—in both cases we might say that, for the Christian, the friction which is produced is of a generative kind. Certainly "academic" history is not meant for all people (and is a somewhat technical affair), for it is not the queen of the sciences, and it is not to be regarded as a substitute for religion or a complete education in itself. Those who promoted its study in former times seemed to value it rather as an additional equipment for people who were presumed to have had their real education elsewhere, their real training in values and in the meaning of life in other fields. Those who complain that it does not provide people with the meaning of life are asking from an academic science more than it can give and are tempting the historian himself to a dangerous form of self-aggrandisement. They have caught heresy from the secular liberals who, having deposed religion, set up scholarship in its place and unduly exalted it, assuming that the historian in particular was fitted to give to his readers an interpretation of life on the earth. "Academic" history would be subject to fewer attacks if our educational system as a whole had not gone adrift, and we had not thrown overboard the very things which are a training in values. In any case, for the Christian, religion comes logically the first, and the study of human vicissitude or of the operations of nature is to be regarded as an additional piece of training.

Yet, if our educationists had been wiser it is not at all clear that Christianity would have been the gainer, or that in the middle of the twentieth century the training for life, the training in values (supposing it to have been efficiently organized) would have been a specifically Christian one. It must be remembered in any case that the real cause of most of the difficulties that trouble the critics of historical scholarship lies in the fact that Christianity no longer reigns or presides over the whole range of our society and civilization. The real problem—and perhaps we are all too slow in adjusting our minds to it—is the initial problem of Christianity in a pagan background, a Christianity which is foolishness to the Greeks, and which can hardly claim from Providence (as though it were a matter of right) that the general influences in the world should not be inimical to it, or that the task of believers should so to speak be made easy for them. It is of course very nice for Churches to have a kind of world in which all the currents of thought are directed by ecclesiastical authority, and all men are brought up so locked in the Christian religion that they are hardly allowed to know that any alternative view of life is even available. It is a question whether such a world could ever be produced, however, save in an intermediate stage in the history of civilization, and after a cruel exercise of force; and it is questionable whether men in the long run would tolerate it, in view of the abuses to which it is liable and the methods to which it is bound to be committed.

In fact what we are faced with in the twentieth century are the disciples of Marx on the one hand, H. G. Wells (shall we say) on the other, the Protestants and the Jesuits, the Fascists and the Liberals, all producing their selections from the complexity of historical facts and their different organizations of the whole narrative of the centuries—all feeling that theirs is the absolute explanation, and longing to see it established as the basis for a universal teaching and examining system. In a cut-throat conflict between these and other systems for the control of schools and universities (in other words for predominance in society) it is not clear that a specifically Christian or Biblical interpretation would in fact prevail at the present day; and though it is a sad thing when any man rejects Christianity, still Christians can hardly have a technical ground of complaint in modern society if universities do not pour all their academic teaching into a specifically Christian mould as in former times. Considering their own record of intolerance and persecution where they had the power, Churches must consider it rather fortunate for them if so often their enemies have been less thorough-going—thankful that, if society is not Christian, it is at any rate not yet wholeheartedly anything else. While we have Marxists and Wellsians, Protestants and Catholics, Whigs and Tories, with their mutually exclusive systems (historical assertion confronted by counter-assertion), many people, confounded by the contradictions, will run thankfully in the last resort to the humbler "academic" historian—to the man who will just try to show what the evidence warrants, and will respect the intricacy and the complexity of events. In the clash of interpretations somebody will sigh in the long run for an answer to the more pedestrian question, the purely historical question: What is the evidence, and what are at any rate the tangible things which demonstrably took place? Men are slow to count their blessings but Christians might even be thankful for this "academic" history at the last stage of the argument—thankful so long as no authoritative interpretation of history and the human drama has been rigorously imposed upon our educational system by an increasingly non-Christian society.

It is true that history—in the academic form that we are discussing—is in a sense very much the study of man. It might be argued—it has indeed been argued—therefore, that any prolonged concern with it must produce or signify a certain "worldly-mindedness," a charge which is no doubt liable to be true in the case of those people who have left their minds unguarded. Those who have used this argument, however, would seem to have forgotten that almost everything else in the world tends to have the same effect, and that most of the human race has to occupy much of its time with very mundane things (especially with human beings) so that the "world" would still have much the same dangers for religion even though we eliminated entirely the pursuit of historical study.

If all this is true then many of the arguments against what is sometimes called the "neutral" character of "academic" history are (in view of the prevailing tendencies to-day) only likely to play into the hands of those Marxist systems which in so many respects embarrass us by their mimicry of the aims and methods of the bygone ecclesiastical order. But more than this, if we consider the shape which the modern world is taking, we might well ask whether it may not turn out that in the long run our "academic" history is more Christian than men have realized.

Christianity and "Scientific" Truth

In the first place it has sometimes been argued that modern science itself took its rise in western Europe (rather than in some other portion of the globe) precisely because of the sway of Christianity in this region. It would require an intimate knowledge of the history of remote parts of the globe and other civilizations than ours to confirm this speculation—a point which ought to soften the attitude of those Christians who disparage historical enquiry, since this is one of the cases where the truths that are reached by such enquiry, though they are limited truths, are clearly relevant to the sort of reasoning that Christians sometimes use. In any event (and considering the advances made in the Mohammedan world in the middle ages) the point was better stated perhaps by Duhem, who thought that it was rather the influence of monotheism which led men in general to the view that the universe is regulated by universal laws. The modern world may even now discover that the religious factor has a place in the history of science, and we may learn that the notion of absolute Truth is not unconnected with religion—the Truth being in point of fact something which you cannot cheat God about. In the case of the early seventeenth century attempts to prove the universe to be a watertight mechanism (Kepler for example), one discovers that religion itself was an original impulse to the endeavour—it was felt that God would be erratic and Creation incomplete if there was a gap or loophole in the mechanistic system. And once in the middle ages, when the Aristotelian philosophy forbade men to think that even God could create a vacuum or an infinite universe, a religious edict (refusing to allow that God should be so disparaged, or that mere human reason should exclude the possibility of certain exercises of divine power) provided a freedom for science which had palpable effects on the history of thought. One is tempted to suggest that there has been perceptible in Marxist propaganda a view of historical (and even scientific) truth which, if it were developed over a long period, might have a very unfortunate effect upon the pursuit of knowledge and the very conception of scholarship. It is difficult to be sure what safeguard there would be against such a "utilitarian" handling of Truth if the world were to go on becoming increasingly pagan and increasingly materialistic in its preoccupations and ends. For this reason is it not unthinkable that in the future Christians should come to find themselves after all the defenders *contra mundum* of a view of Truth to which their idea of God is not irrelevant. They may find that it is they who after all must stand as the defenders of that whole intellectual universe in which the academic historian's search for truth had meaning and could be pursued with the required austerity. The ideals of "academic" history may transpire to have been the legacy of a Christian civilization after all.

We might have to confess that some Marxists in the present have been better and some Christians in the past much worse than the above argument might seem to suggest, so that such an argument is somewhat speculative and no doubt open to challenge. We are on firmer ground in another field, however, where we may think of history as an extension of our personal experience, a relationship with a wider range of human beings, a bond between us and all the generations that lived before us. Here a point arises which is a technical one for the historical student but which touches on Christian maxims concerning our way of living in the world. There may be some justice in the claim that history is a "science," but if so it is a science dominated by the fact that its particular kind of truth can only be attained by imaginative self-giving in human sympathy.

Why History Is Not "Neutral"

In a pagan world it is possible that vast hordes of men—great nations—each utterly sincere in its blind self-righteousness, may hate and fight, not even knowing what sympathy and charity are able to do in the way of creating bridges between people who are not like-minded. Similarly it has always been possible to have a successful religion or a successful regime which in the same blindness of self-righteous pride would build its whole version of history on hatred—on the sheer denunciation of its predecessors. In this latter case a mitigation will come especially when there has been a lapse of time, but it comes because more charitable souls have a desire to understand even the despised and defeated, even the people towards whom they were not initially disposed. And historical understanding depends upon this compassion, this urge to see even the outcasts as human beings who could fall in love or be hurt—as people who "if you prick them they will bleed." If some objector thinks that all was very well in the first instance or that, with a wilfully Protestant view of the Reformation and a wilfully Catholic view, all that is necessary is to add the two together or take the average, such a person is even technically wrong in his idea of the nature of history; for the man who has wider charity and imaginative sympathy, and seeks from a higher altitude to embrace both parties in his comprehension, not only shows greater compassion but uncovers a further range of truths about the Reformation itself. Even of the worst of men, even of Hitler, the historian will want to know how a lump of human nature—or rather a schoolboy playing in a field—could ever have come to be like that. History, as it rises above mere militant partisanship—always committed in a world like ours to loving sinners in spite of the sin—knits the broken threads of time together again, and, brooding more in sorrow than in anger over the conflicts that have so often torn the human race, brings to them human understanding and a reconciling mind. In a world that is fast losing these arts that might have saved it, the practice of this kind of historical study is a useful training for the actual conduct of life; and to Christians who believe that the world needs charity such exercises reveal new paths for the sympathetic imagination—expose the ways in which we have been too narrowly uncomprehending in the past. Such history is not "neutral" save in the sense that it seeks to extend its charity to all men, and if some history lacks the warmth and sympathy of this ideal we must not condemn academic history as such but merely recall it to its own essential principles.

History and Personality

All this implies a high estimation of human personality regarded as (from a mundane point of view) an end in itself. Such a valuation may owe less to Christianity—less even to Greece and Rome—than some people assert, for in some of the aspects of it which the world most prizes it has seemed to come with an advanced and highly differentiated civilization, even while the power of religion in society was declining. In the long run it is always of great significance, however, that Christianity sees human beings as souls meant for eternity, unlike anything else in creation; and it is possible that the greatest of future conflicts between Christian and pagan (or at least the greatest of those which have relation to mundane affairs) will concern the question whether the individual personality is regarded as an absolute or man is envisaged as merely part of nature, part of the animal kingdom. Now the thing which we have come to regard as history would disappear if students of the past ceased to regard the world of men as a thing apart—ceased to envisage a world of human relations set up against nature and the animal kingdom. It has been said that if a lamb should die in May, before it had reproduced itself, or contributed to the development of the species, or provided a fleece for the market, still the fact that it frisked and frolicked in the spring was in one sense an end in itself, and in another sense a thing that tended to the glory of God. This view would serve to typify the attitude of the historian, as distinct from that of the biologist, only interested in such history as relates to the development of the species as a whole. Because in man the spiritual and the temporal intersect, every moment, every individual, matters, every human soul is worth befriending, and those who laugh at a research-student for burying himself in a period of the Icelandic past are overlooking the fact that human life always has its interest—they are somewhat like the people who laugh at the idea of a God who notices the fall of a sparrow. Furthermore, to the historian every event, every action, is to be studied not as an external thing, not as a mere part in some system of mechanics, but as a thing incomprehensible save as it goes into or comes out of human minds; so that de-personalized studies, mere schematizations—pure diagrams of social interaction and sociological development—contravene the real nature of history precisely at the point at which they offend the Christian view of life.

If the academic historian lays out the story of the centuries—a story which in a certain sense is the extension over long periods of what we see of life in the few decades of our actual experience—the question next arises: What happens to this story, this panorama of world-history when one believes in Biblical interpretations and in the Christian view of the whole drama of human life in time?

II

The Christian and the Biblical
Interpretation of History

It is to be doubted whether men really decide their religion by meditations on astronomy or geology or the structure of matter, even when they imagine that they are being so "scientific" in their approach to the problem of their destiny. The conclusions that they draw even from the discoveries of the natural scientists depend upon a more fundamental factor of which they may be hardly conscious—namely the posture they themselves have initially adopted towards the universe. In a similar way our interpretation of the whole drama of human life on the earth is, paradoxically enough, hardly to be decided by merely envisaging history itself and gazing upon the course of the centuries. The crucial decisions are the ones that we take when we adopt our attitude to life itself, and they belong to the most intimate regions of our personal and private experience. Those who believe that it is God whom they have known in the solitude and the profound silences will also know him in history, but the converse process —seeing God first of all in the ordinary course of secular history— is not an easy thing to imagine taking place. And if both these processes are combined in the New Testament, so that we can say that God is revealed to men through Christ in history, we must remember that he is not revealed to the more external ob-server who puts on the thinking-cap of the ordinary historical student. He is revealed only to those who take the Gospel home to themselves in that innermost region of the personality where the most intimate of our decisions are made. Indeed, history of any sort has hardly been properly assimilated by us until it has been knit into a single fabric continuous with our own internal life and experience.

We are all familiar with the practice of taking the various kinds of moral precept or spiritual truth contained in the Bible and transposing them out of their original context so that we may appropriate them to our own daily lives and labours or vicissi-tudes. If such a transposition were not regarded as admissible the Bible itself would hardly be a relevant book for the people of the twentieth century. On the same principles, and in accordance with the rules that govern the use of the Bible in our private experience or our devotional life, we are bound to take over a Biblical interpretation of history—indeed those who rejected such a thing could hardly escape being in the position of rejecting the Bible itself. Not only are our private affairs and intimate experi-ences part of the web of universal history, but such an injunction as "Fret not thyself because of evil-doers," however we may inter-pret it, must apply to public and national events as well as to the private vicissitudes of men. And as the Old Testament is so often concerned with the Jewish people as a body, rather than with in-dividuals as such, it is not easy to see how readers could find it valuable in their religious experience if they did not discover in it an interpretation of human history—one which, if not fully the Christian interpretation, is at any rate valid so far as it goes. If some people object that this can hardly be the case since the events narrated in the Gospels changed the character and altered the texture of world-history for all the future, it must be answered that whatever truth this thesis may possess on a final analysis it does not destroy the validity of Old Testament teaching at its own level, that is to say, at the particular points where that teaching happens to attack the problems of history.

The Character of Old Testament History

It is possible that the power of much of the Old Testament teaching about history would be more vividly appreciated, and its relevance to the twentieth century more readily recognized, if only we could rid ourselves of an obsession and genuinely convince ourselves that the history of the ancient Hebrews was fundamentally of the same texture as our own. There is ample evidence that in their own great days, in the age of the mighty prophets for example, they looked back upon their own distant past in the way in which we ourselves now look back to *them*; and in manifold ways they express the thought of Psalm 44, "We have heard with our ears, O God, and our fathers have told us, what work thou didst in their days, in the times of old." It would appear to have been one of the functions of the great prophets to point out that God was still acting and intervening in history as in the time of Moses—that history in their more modern age and the history of the days of their forefathers was all of one piece. And there is ample evidence of the repeated failure of the prophets to achieve the task—ample evidence of the desire that God should show himself more plainly, as in the ancient days, so that people should not be able to ignore his part in history any more.

What was unique about the ancient Hebrews was their historiography rather than their history—the fact that their finer spirits saw the hand of God in events, ultimately realizing that if they were the Chosen People they were chosen for the purpose of transmitting that discovery to all the other nations. Their historiography was unique also in that it ascribed the successes of Israel not to virtue but to the favour of God; and instead of narrating the glories or demonstrating the righteousness of the nation, it denounced the infidelity of the people, denounced it not as an occasional thing but as the constant feature of the nation's history, even proclaiming at times that the sins of Israel were worse, and their hearts more hardened against the light, than those of the other nations around them. The great religious thought which stands as the Old Testament interpretation of the whole human drama was clearly the work of a few select souls—of great prophets often standing with their backs to the wall, for example—in a nation whose history otherwise ran under very much the same rules as the history of other peoples. It is even possibly true to say also that the makers of the Old Testament, while having an extraordinary feeling for the might and grandeur of the human drama, were not historically-minded in the sense that this term has come to have in the twentieth-century—not interested in seeing that the past should be accurately recorded for its own sake, or that all the great episodes in the history of their country should be put into narrative for the sake of posterity.

It is remarkable that so small a nation should have come to occupy so great a place in the history of the world; and George Adam Smith, in his *Historical Geography of the Holy Land*, has given us cogent reasons for not accepting any mere geographical determinism as the explanation of their peculiar historical destiny —cogent reasons for regarding the story as an example of the triumph of the human soul over physical conditions. It was a stormy history that the country had, moreover, with only a remarkably short period of political independence, and it has been questioned whether any area of the earth felt the tramp of troops more often than Palestine did in the period down to the opening of the Christian era. The great original contributions of the ancient Hebrews to both religious and historical thought are curiously connected with the period when this stormy history came to its climax, and the country was engulfed in the conflicts between the vast empires in their neighborhood. Once again it is necessary to remember that their fate in this respect was not unique—they experienced that cataclysmic history which we find constantly recurring as the centuries succeed one another in this precarious universe. It is we who have long been spoiled by a feeling of security and a dream of ordered progress, so that we have forgotten the very nature of history. It is only now, in the middle of the twentieth century, that we are beginning to know once again what this cataclysmic history feels like—know something also of the moral paradoxes with which it confronts us, and which so often confound those who are groping for some "meaning" in the human drama. The teachers of Israel, realizing the relationship between Jehovah and the nation's history, and conceiving that relationship in ethical terms, set out to vindicate the moral character of history at the critical point where the difficulties are at their maximum—the point where men (as we have seen even in twentieth-century Europe) so often feel themselves victims of blind chance or sink into fatalism.

So it might be claimed that for all students who hope to understand either history or the problem of its interpretation the importance of this ancient literature is greater than is usually recognised. It would not be an exaggeration to say that those people who study merely nineteenth-century history, and see the nineteenth century running by apparently natural processes into the world of the present day, are liable to fall into a routine kind of thinking which actually incapacitates them for any appreciation of the profounder characteristics of our time. In the ancient world, where a long series of centuries allows us to see how historical episodes ultimately worked themselves out; in more simple forms of society where events are less entangled, so that causes may be seen more clearly leading to their effects; in antique city-states, where we can more easily view the body politic as a whole and where developments are telescoped into a shorter compass, so that the processes are more easily traceable—in all such cases as these the student of history may reach a profounder wisdom than can come from any vision of the nineteenth century through the eyes of the twentieth. Even if this were not true it might be well if all historical students were induced to occupy themselves with an internal analysis of a few mighty episodes in history—the fall of the Roman Empire, for example, or the scientific revolution of the seventeenth century—episodes which have represented the climax of human vicissitude and endeavour, high peaks in the experience of humanity on the earth. By all these lines of argument the events in the centre of which stands the famous Exile of the ancient Jews ought to be a part of the curriculum of every serious student of the past. They are more contemporary with the moral predicament of this part of the world since 1939 or 1945 than anything in the history of the nineteenth century. And they enable us to see to what an extent our religious thought itself has developed from wrestlings with God and reflection on tragic history.

God in History

On the Old Testament view God is intimately involved in all the events of history, and the twists and turns of the resulting story seem as though they had been shaped by a person—whatever men may do or try to do, the total result (as the historian would synthesize it) shows the effects of his hand. In this he continues the original work of creation, with still fresh devices at his command, still fresh surprises in reserve. "I have shewed new things from this time. . . . They are created now and not from of old; and before this day thou heardest them not; lest thou shouldest say, Behold, I knew them." His policies wait upon the conduct of human beings, however—now he "hath a controversy with the nations, he will plead with all flesh"—now he says, "If that nation, against whom I have pronounced, turn from their evil, I will repent of the evil I thought to do unto them"—so there is always great elasticity and variety in the developments that the story may take at any given moment. While men are about their several purposes there is a history-making going on over their heads to compound the results of their efforts and turn them into a pattern. And, if they had done their thinking on this basis, men would hardly have made the mistake of assuming that since the nineteenth century was of a certain character, the twentieth century would go the same way, only more so. Furthermore, this Providence, working upon the lives of creatures who are still free, has the power—a power so necessary for the salvaging of human history—of turning evil into good. As Joseph said to his brethren, "As for you, ye thought evil against me; but God meant it unto good, to bring to pass, as it is this day, to save much people alive." In fact, the hazardous moments for the world must be those when God merely withdraws his hand, merely hides his face from men, merely allows events to work themselves out. There were times when it was an additional sin of the Jews to resist events that seemed an ordinance of God—to fight against Providence—as when under the shadow of the terrible empires of Assyria and Babylon Isaiah had to protest against any resort to an alliance with Egypt, and Jeremiah preached, not resistance, but "collaboration," to the horror of the jingoists. In such cases the refusal to listen to the word of God through the mouth of the prophets brought its own automatic retribution.

Judgment in History

I do not know if it would be rash to suggest that, as our appropriation of religion must differ in some respect from that of the ancient Jews, this view of the intimate relations between God, morality and history will be likely to find favour only with those who have previously been convinced of the operation of God in their personal life. In other words our interpretation of the human drama is an extension of the interpretation that we give to life in our personal experience of it. Granted so much, it must be difficult for the religious mind not to feel that the setbacks, the adversities and the sufferings of life are (according to circumstances perhaps) either a judgment on one's sins, or a discipline for the soul, or a testing of character, or any combination of these.

Once again it must be true that if God works upon our lives in detail and in the intimate things, he must operate generally in history—we can move on to a broader canvas and equally say that there is "judgment" in history for example. Historians have long recognized the fact—recognized that a Napoleon and a Hitler came to a downfall which was a retribution for their sins and excesses. Even when historians seem to speak a different language we ought to be careful not to be deceived by what is their technical form of phraseology—a phraseology that is adapted to their particular purposes. They may say that the capitalistic system has been carrying us to ruin for a generation, but they know that they mean that it is human cupidity (not sufficiently checked under this so-called capitalistic system) that has been over-reaching itself. They may say that absolute monarchy was once the evil, but they mean in reality that human beings are always unfit to be allowed to hold unlimited power.

Sometimes the Old Testament writers seemed to regard the judgment as implicit in events, or embedded in the very constitution of things. Jeremiah says, "Thine own wickedness shall correct thee"; "Your sins have withholden good things from you"; "Behold, I will bring evil upon this people, even the fruit of their thoughts"; and "Do they provoke me to anger? saith the Lord: do they not provoke themselves to the confusion of their own faces?" Later, in the *Wisdom of Solomon*, we read: "For the creation, ministering to thee its maker, straineth its force against the unrighteous for punishment." Sometimes God has only to withhold his protection—"I will hide my face from them, I will see what their end shall be"—and the terrible penalty comes from his non-intervention. When the loom of mighty empires gives presage of colossal disaster, the great prophets see God himself coming with judgment, however, and ask themselves "What are the sins of Israel?" They did not—as the twentieth century would have done—merely cry out in self-righteousness against the sins and aggressions of Assyria.

Messianic Ideas

The Old Testament view of history comes to us in stages, but at each new stage in the development the old is not superseded—it forms the substratum for the new, so that layer is piled on layer, and the more ancient ideas may come to appear different in a new context. At the climax of the story it is difficult to see how anybody could understand an "historical religion"—or accept a Christian view of history—without this initial substratum, this Old Testament interpretation of secular history. Even the thunderous prophetic message of judgment in history, though at times it almost seemed to denote a final judgment, an utter destruction, was soon realized not to have superseded the kind of history based on the Promise—it was superimposed upon that history but it did not cancel the Promise. When disaster had come and despair was at its worst this was remembered and there appeared that interesting historical conception, the idea of the Remnant of Israel, who, keeping faithful or chastened by their sufferings, should return to the homeland and still inherit all the fulness of the Promise. Consonant with this, there emerged the messianic hope, which issued on occasion in such fine nostalgic poetry, and which seemed to promise the ancient Jews a great political future, still, amongst the nations of the earth. Some writers have asserted that this political messianism was an important contribution to our interpretation of the human drama, since it showed history moving to a purpose, moving to an end that should lie within history and take place on the earth itself. It has even been suggested that the modern idea of progress issued in reality out of this phase of Jewish historical thought. It is not clear, however, that this political messianism did not lead the Jewish people to political mistakes and disasters, even if it merits some of the praise it has received. We pay it a doubtful compliment in any case if we suggest that it may be the parent of the modern idea of progress. Certain periods of strain in the twentieth century have issued in forms of political messianism which could only add to our misgivings on this point. Here is a branch of Jewish thought on the subject of the human drama which was to be wonderfully redeemed at a later date and was to become wonderfully relevant, when it became spiritualized and was brought to a different kind of fulfilment in the Christian revelation. Moreover, before the opening of the Christian era, apocalyptic thought had attained the view that only a direct intervention on the part of God could rescue man-in-history.

The Problem of Undeserved Catastrophe

A view of history which sought to show that events had a moral character and that all events can be turned to moral benefit, was bound to find its crucial challenge when it had to consider the incidence of suffering in the world. Ezekiel was evidently puzzled to see some of the unrighteous surviving the catastrophe that God had brought upon the wicked, and he seems to have conjectured that they were allowed to live in order that their wickedness should provide standing evidence of the reason for the Divine intervention that had taken place. "Ye shall see their ways and their doings: and ye shall be comforted concerning the evil that I have brought upon Jerusalem . . . and ye shall know that I have not done without cause all that I have done." The converse case—the case of the innocent sufferer, a case where a doctrine of "judgment" was palpably insufficient, whether the sufferer were a nation or an individual—exercised some of the highest thought of the Old Testament. Once again the historical problem was at the same time a problem intimately personal, bound to be affected by one's ideas on the whole problem of suffering. The Old Testament had to meet the problem of cataclysmic history in its extremest form, for they lacked the doctrine of original sin, which alters the bearings of many things in history, and they lacked the doctrine of a future life and a judgment after death, with a consequent re-distribution of fortunes in another world. The picture of the Suffering Servant (e.g. in Isaiah 53) voluntarily accepting his tragedy as one who bears the sins of others—as part of his mission in the world—is the highest point reached by the ancient Jews in their attempt to comprehend (and to reconcile themselves with) their national tragedy. It provides a pattern of what takes place in history, what helps to redeem history, what reached its perfect fulfilment in the life of Christ—namely suffering accepted as voluntary and as vicarious, and regarded as the only way of reconciling history with morality or reconciling oneself with history. In other words, we interpret human history and we make our peace with it, not as students who merely observe and calculate, not as God who sees and comprehends all, but rather as participators in it who have discovered for ourselves the rôle which brings harmony.

12

The Christian and the Marxian
Interpretation of History

It is necessary to start from the fact that (so far as the technical historian's universe is concerned) it is human beings who make history, for their vitality keeps all the wheels in motion, and ideas exist because human beings have brains. At the second stage of the argument, however, we realize that the initial appearance of "sovereignty" in self-acting human beings is to some degree an optical illusion. Men make history, it is true—they do not merely sit and suffer it—but they are partly made by history in the first place, gravely affected by the fact that they are born into a certain social class, or brought up in a Protestant household or taught to do their thinking in the English language. The problems they have to face, the very predicaments in which they find themselves, are also the products of history—things which the historian would try to explain by examining their antecedents. So all men are entangled in which might almost be called a system of necessity, they are caught in a complicated piece of network which conditions their actions at multitudinous points and with varying degrees of constraint; though it must be remembered that they are real live creatures entangled in the network—not helpless prisoners, fastened or inert within it.

There is a particular sense in which it may be said that the academic historian is interested in the action of men precisely in this latter aspect—that is to say, in its aspect as conditioned action. We may regard Charles I as a free man, who might have acted differently, but the historian can hardly reach deep enough to measure the man's responsibility or to do more than charge him with his share in our universal sin. The historian deals rather with what we might call the underside of the tapestry, the discoverable roots of human conduct. He takes Charles I's action as it was in fact committed, and assembles around it everything that he can gather concerning that monarch's position, his physical structure, his education, his motives, his temptations, and the like; in other words, he seeks what we call the "historical explanation"—which must never be regarded as the total explanation—of the episode in question. For this reason the historian (when he is analysing, and not merely narrating) is a student of that necessity which conditions, though it must never be regarded as determining, human action. And an interpretation of history (in the sense that Marxism is an interpretation) is a commentary on that system of necessity—a commentary on the providential order.

In this sense we need not be dismayed by a "materialist" interpretation of history; indeed we might say that technical history as it is usually understood is just the materialist interpretation of the course of ages. If we remember our original thesis, that it is human beings who make history, and keep in mind a second thesis —namely that nothing less than the whole of the past is necessary to provide the "historical explanation" of the present—these two things set definite limits to the scope and import of any interpretation of history whatsoever. I must say, then, that if a few blows on the head with a hammer may be sufficient to put me out of the historian's universe altogether, I ought not to be surprised if an interpretation of history, of the system of necessity (in other words of this aspect of the providential order), should have a hard and materialist quality. I need not feel that an economic interpre-

tation of history is an absurdity, for it may register the fact that
certain economic conditioning circumstances may set insur-
mountable checks on human action or barriers to human achievement. Such a view would go further than would be legitimate in
any interpretation of history (in the academic sense that we are
considering the word) if it pretended to assert that man can live
by bread alone; but it bases itself on the thesis that in the technical historian's universe man, however spiritual he may be, cannot
live without bread. It may be true that no great advance in literary culture could take place even so long as men were living in
the constant unmitigated fear of starvation. We ought to be prepared to consider that perhaps there could be no advanced science, no abstruse philosophy, no modern rationalism until the development of means of production had brought about a certain
progress in society, complicating the social structure and producing a division of labour. It is in relation to this particular line of
thought that we have to consider the merits and the defects of
Marxist history.

Lord Acton thought it the duty of the historian to understand
the enemy better than the enemy understood himself; and it
could hardly be denied that this is more particularly the duty
of the Christian—most of all perhaps in dealing with Marxism,
where we shall never diagnose the evil until we have discovered
the good. Nothing should be further from our minds than the
easy game of making debating points out of incidental issues,
while evading the labour of disengaging that element of truth
which all systems must possess before they can gain the slightest
grip on any great sections of opinion. If politics had never come
into the question it would long ago have been recognized more
frankly that Marx and Engels, in spite of the weaknesses which
seem so obvious, made an important contribution to the study of
history. We make a mistake sometimes when we compare our
history as it exists in the ideal with Marxist history as it exists in
actuality—we have in mind a rarefied form of academic history

which in reality has only a limited range of effectiveness in our world, and we overlook the fact that Marxist history may still have its points against the kind of history which the average Englishman holds in his head. Also it is difficult for people to change the accustomed framework of their history, and it entails a greater uprooting in the mind than most men are capable of; so that the Marxian presentation of the story revolts many people as a mis-shapen structure, because it confronts them with something so different from the scheme of history with which they are familiar. So we are able to bring ourselves to the fact that contemporary Communism, precisely by reason of its interpretation of history, is one of the most formidable competitors that Christianity ever had to face.

The Marxists did not invent but have given currency to that dialectical method which tends to see a conflict in society—thesis fighting antithesis—until the emergence of the new thing, the synthesis, puts an end to the old conflict under a system that both embraces and supersedes the earlier parties. This formula, whatever its weaknesses, is capable of supreme elasticity, while at any rate it saves people from a more unfortunate error that is still common amongst us—the habit of seeing in history a more simple progress, an ascent in a straight line of logical development. Those who say that Protestantism led to toleration and then so easily assume that there was something in the original form of Protestantism which logically led to modern liberty, have a less satisfactory pattern in their minds than those who see the conflict of Catholic and Protestant leading to a toleration which was initially not deducible from either of the two. The Marxist is more formidable in one respect as a result of this: he is the more likely to have his mind prepared for the emergence of surprises at the next turn in history.

It might be asked why we should ever need an interpretation of history (in the sense that we are now considering such a thing) —why we should not merely historically enquire. It appears, however, that people may survey the evidence and get the facts of a matter correct, while telling the story, so to speak in the wrong universe; and Marx did some good by inverting the policy of his "idealist" predecessors, and by reminding us that human history is curiously fastened to the earth. When Napoleon said that the nature of weapons decides everything else in the art and organization of war; when Parry once showed how the nature of instruments conditioned a development in music at an important point; when Holland Rose described how men opened their minds quickly enough for the great voyages of discovery once the requisite nautical apparatus was at their service; when Virginia Woolf gave £500 a year and a room of one's own as a necessary condition for the production of literature, they may all of them have been wrong, or they may have been exaggerating, but they possessed the right kind of feeling as historians. There may be no known limits to the potentialities of the human mind, but so long as no instrument exists except a tin whistle certain limits are set to the kind of music which in fact will be composed in the world. And when, at the opposite extreme, historians attribute the break-up of feudalism to the decline of the feudal idea, or account for the end of the tribal system by talking of the decline of the tribal idea—when they speak as though the human mind just decided to arise and expand at the Renaissance—they are reading their history in the wrong universe and failing to hold the story down to earth.

It is perhaps the Marxists more than anybody else who have taught us to see our history structurally—not to be content with the mere surface-drama of events but, when we are studying an episode, to make a kind of geological examination of the ground underneath. Even if we are interested primarily in the history of ideas, we should not today study the pure development of ideas in the air, but should hunt out their relationships with the social environment and with the social changes that had been taking place. The Marxists are not original when they assert the fundamental character of the class-conflict in history, but again they are responsible for much of the currency which this thesis has attained; and I think that the modern historian would hardly study, say, the spread of the Reformation without examining the structure of society and the tensions between classes in the region concerned. When the Marxists say that the ultimate source of significant change is to be found in an alteration in the means of production, they make an assertion which is difficult to prove or to disprove, especially as their system is capable of extraordinary elasticity. They may have in mind chiefly long periods of history; for example the differences between feudal society and the modern bourgeois world—differences which they see running through the whole fabric of the resulting civilization. They would seem to make a mistake, however, when studying history (where everything so quickly becomes entangled with everything else) to imagine that things can be easily unravelled, and that one can lay hold of a single factor that is the ultimate factor, a cause that lies behind all other causes, an irreducible "starting-point of historical change."

Partly owing to crudities and perversions; partly owing to an entanglement in ideas or prejudices which happened to be current when the system was being developed over a century ago; partly owing to the humble character of many of the people who have followed and extended the system, and who fell into some of the pitfalls of self-taught history; partly again owing to partisanship and polemical intention—in other words, to the fact that the system has been identified with a political cause; and finally, because its devotees have so often shown rigidity in respect of principles of their own which required supreme elasticity of mind—for all these reasons the men who have actually written Marxist history have tended to produce a kind of history and a type of historical universe with which we must quarrel very radically indeed. If they have been right in calling attention to features of the historical process which had been ignored or insufficiently appreciated, they have fallen into errors parallel to the ones committed by their predecessors. They have seized upon one aspect of the truth and assumed it to be the whole.

In a fundamental sense—in a sense which reaches out far beyond the mere question of an historical method—the Marxian outlook takes materialism for granted, and takes it as the whole of the story, assuming that the concrete material things are sufficient to account for everything, and that historical explanation is a total and self-sufficient explanation of the universe. The Marxian theses concerning the processes of history could be held in a certain manner by those who rejected this philosophy as a whole —just as one might believe in the class-less society without necessarily being an atheist. The idea that we should be able to explain a thing if we could tell its history, the mundane self-sufficiency of the universe, and the denial of God are things not peculiar to Marxism, in any case; but were rather taken over by it from the secular liberals.

The Marxist historical method becomes a grave danger, however, to those who hold this general materialist outlook; and the danger is increased by the fact that in the Marxists there is "materialism" in a more popular sense of the word—an approach to history which is too greatly preoccupied with the question of the distribution of worldly goods. Even if the cause of this is a certain altruism and a charitable intent, the kind of history which results is liable to be possessed with a kind of "materialism" that is actually corrupting to the mind. In so far as they contain important elements of truth the Marxian principles would be useful to anybody rather than a Marxist—most useful of all perhaps for a secure Christian, proof against the charm of materialism as such, yet anxious to keep in touch with the hard earth. It generally happens that the Marxists are too materialistic in their view of the nature of human beings, and, if in a certain sense they have done service by tearing this mask from human nature, they have done great harm by their lack of interest in human beings as such, their lack of awareness of that universe which lies inside a personality. In reality amongst the Marxist historians this general attitude (though it is not by any means a necessary consequence of the Marxian theses concerning the historical process) has a great effect upon the texture of the narrative that is produced and the nature of the universe that is being described. Economic self-interest too often becomes not merely a bias inside human beings and a fundamental factor in history, but the very motor of men's activity and the standing subject of their mental calculations.

The actual processes of history are distorted through this error to a less degree than some people might assume. If a man is not interested in promoting his business, or even if he withdraws from business in order to pursue one of the arts, his place in the story will be taken by other people who do have the push—so that the course of economic history will run on much the same. Also it is not always realized how greatly a man born in a certain social class, even if he is charitable and well-intentioned, will be constricted in his vision because he has been accustomed to viewing events from the platform of that class. The statistics of a vote taken in this country at the present day on the place to be given to the classics in our educational system might very well afford an example of this.

We have seen that human action is not really "sovereign" action and we have noted how intent the historian must be for those discoverable factors which condition the conduct and even the thoughts of men. All the most dangerous assumptions in the world have been smuggled into the argument, however, if we think that these factors do more than "condition" human conduct in varying degrees and if we accept them as real "causes," real makers of history. It may be true on the one hand that, so long as there exists no better instrument in the world than a wooden pipe, a serious limitation is bound to restrict the kind of music which it is possible to produce. In such a case, however, the frontiers that are set to any enterprise of the mind, like the fencing round a house, may block a lateral expansion on any side, while still leaving room to soar, still leaving everything up to heaven wide open for the creative activity of human beings. Those who see in history a sort of continuous in-breeding, on the ground that the men who decide the course of things at any given time are themselves only the product of their age, make a grave mistake by the introduction of this word "only." They forget how all the influences and ingredients which the age and environment may provide are liable to be churned over afresh inside any human personality, so that each man is a separate fountain of action, unpredictable and for ever producing new things— each man is a possible starting-point for historical change. The higher we go in the order of things, the more this becomes true. In some of our calculations we make too little allowance for the effect of conditioning circumstances especially on other people; but perhaps it is the neglect of the spiritual character of man which makes human beings also, very often, forget their liberty. It is not clear that anything (unless it might be the very condition of unpropitiousness) was propitious for a religious revival in the eighteenth century except the emergence of John Wesley.

Having asserted the fundamental and structural importance of the economic factor in history, the Marxist historians, when they write their narratives or studies, too often seem to assume that this is the only thing which matters, or that all the other things in human history may thereafter be taken as read. A caricature of this heresy is provided by the youth who imagined that a knowledge of the basic conditions of the Elizabethan era would provide one with a formula or a condensation of the culture of that period which would render the reading of the more incidental manifestations of that culture (Shapespeare himself, for example) superfluous. This again is the effect of materialism in the popular sense of the word and it would not seem to be correct to regard it as an inevitable consequence of the Marxian method as such. If the Marxist may be right when he puts the economic substructure at the bottom, he is not permitted to place it also at the top, or alternatively, to dismiss everything else—art, the constructions of the intellect, the achievements of personality, and the spiritual things—as *mere* superstructure and therefore unimportant. It is one thing to recognize the significance of economic factors in history, or a kind of finality which they may even possess "in the last resort." It is another thing entirely to see the history of religion or culture or even politics as almost a crude by-product of economic history.

The very faults which are responsible for what might be called the ugliness or repulsiveness of Marxist history, are separable from the essential Marxian theses concerning the historical process; and a great gulf exists between the extremely subtle theories of history which the more high-brow Marxists have put forward and the coarser kind of history which Marxist historians generally seem to produce. The faults of the system are often capable of correction within the system itself, and it is curious to note that the Soviet historian Pokrovsky argued against excessive resort to the crude economic interpretation of history, while Stalin a decade later decreed that Pokrovsky had gone too far in this direct economic interpretation himself. It took centuries to refine out of the traditional English or "Whig" interpretation of history the crudities (astonishingly parallel to the crudities of Marxism in what is still its youth) which it possessed early in the seventeenth century. There is evidence that Marxist history in its higher reaches will be capable of issuing in research of an academic character not less imposing than the research of some of those people who think that they are writing history without any presuppositions at all. If people point to actual deceit or prejudice or self-delusion in Soviet historical writing, much of this is not organic to the Marxian method as such, of course; it is the too common characteristic of any history that is written in support of a government or in praise of a regime, or in flattery of one's country or in the service of the state. Many of the people who think that they have answered the Marxian view of the historical process as such seem like so many of those who imagine that they have answered Christianity when they have only answered some mistaken vulgarizations or popular errors. The danger lies in the fact that a number of Marxian theses concerning the way in which certain concrete things are connected with one another in history make it easy to smuggle into the story a materialism which is sometimes gross; and it may be a subtle matter to expose the conjuring-tricks in detail.

13
The Christian and the Ecclesiastical
Interpretation of History

It would appear to be the case that many people would rest the historical defence of Christianity not so much upon the general religious interpretation of the human drama as a whole, but on the virtues of that period when the religion was universally accepted and the Church presided in a particular way over the destinies of our part of the world. Some people have even imagined that the politico-ecclesiastical synthesis of the middle ages was the norm or the ideal from which the West has departed in a comparatively recent period of aberration. One of the great issues which Christians might do well to consider is the question whether that medieval system is to be regarded as a norm or even as an experiment that is in any sense repeatable; or whether the normal situation to expect in the world is not rather that of the Church in the early centuries, before the Roman imperial government had been won over—namely the situation of Christianity against a pagan background. Those who tend to produce an historical defence of Christianity by reference to the virtues of that system which prevailed in the medieval and early modern times, are committed to a particular kind of defence of the Church in history—a defence of its mundane achievements and its establish-

ment of what purported to be a society based on Christian principles. They impart, therefore, what is really an ecclesiastical fla-
vour to their interpretation of general history as a whole.

The issue would present itself more vividly to us if we were to
imagine either communism or Christianity capturing a great section of our globe, and establishing its New Jerusalem. It is easy
to see that once the victory had been achieved by some such
system constant persecution or coercion might not be necessary
to prolong its power from decade to decade, even perhaps from
century to century, in the after-period. The orthodoxy that once
succeeded in imposing itself might continue its dominion by virtue of habit and intellectual indolence, or through its control of
education, or by the very exhilaration of its success. It is when
we are contemplating the manner in which such systems originally establish their sway that we meet a paradox which should
be a serious matter for contemplation. The communist system, in
much of its method, its missionary policy, its thirst for orthodoxy,
its authoritarianism, and its severities (when severities are needed
for the consolidation of its power)—above all perhaps in the papal
insistence that all kindred systems should be in communion with
Moscow, the new Rome—is almost frightening in its mimicry of
the Christian Church at certain periods in European history.

One of the most moving stories that history has to offer us—one
of the clearest cases ever known of the meek inheriting the earth—
is the spread of Christianity, from a narrow and unpromising
home, over the length and breadth of the Roman Empire. In the
very earliest centuries of the Church Christians would hardly be
accused of having made it their object to capture the Roman
Government. This was a thing which was added unto them—a
gift of Providence to them for the time being—and an enemy
might make a severe criticism of the way their successors used the
gift, though an historian, aware of the weaknesses of human nature generally, would find it understandable. A general unanimity
in the Christian faith—or indeed in anything else—is a solemn and

awful thing, not to be counted as ordinarily achievable in adult states of society without the resort to methods that are grim to contemplate. Even if we ask how it was achieved in barbarian Europe after the fall of Rome, some inspiring examples of missionary work and martyrdom can hardly blind us to the degree to which our religion was often imposed from above, sometimes after terrible warfare, sometimes as an accompaniment of cruel military conquest. And we must remember that, when the unanimity breaks—whether we imagine it a Christian or a communist orthodoxy which is losing its hold upon a people that has been attached to it—it is the Christian precedent which justifies us in the view that terrible measures will be resorted to by the threatened system in the endeavour to keep its control.

There is something in the nature of historical science which makes that study adequate to the examination of the changes of things which change, but inadequate for the recording of the permanence of things which stand. History can more easily describe the stages in the expansion of Christianity than measure the influence of that religion as a standing factor in European history. Apart from this, history easily allows many human things—like men's falling in love—to evaporate out of the story, though they may be of overwhelming importance in actual life. In particular it easily fails to seize hold of the spiritual side of religion, because here is something that tends to elude the historian's peculiar kind of net. Religious history is easily transformed into a form of politico-ecclesiastical history, and it comes to be imagined that the Church is to be justified by the historian in a way that a political body or a utilitarian corporation is to be justified. People forget that the grand fact of European history is the constant preaching of the gospel, the conversion of souls to a more authentic appropriation of religion, and the ministering to their spiritual necessities—a thing which the historian can hardly go on repeating, though it is the same year after year and generation after generation. Here is the impressive story, the point where the work of the Church has eternal validity; here is the task in which it has never failed, the task in which it must not be thought to have failed even when many men are turning their backs upon religion. We often do an injustice both to Churches and to Churchmen because we hold a skeleton of politico-ecclesiastical history in our minds and fail to picture the life of the clergy in its intimacy and its fullness.

In a parallel manner those who seek to measure the importance of Christianity in the history of European civilization must be prepared to deal with something much more subtle and delicate than the kind of politico-ecclesiastical history which they may find depicted in its larger lines in ordinary text-books. The problem is by no means an easy one for we cannot unthink our European history and imagine what it would have been like without Christianity. We may even be deceived if we base our inferences on an inversion of what we see when we observe the changing character of the world as it has become increasingly pagan in our time. The part which Christianity played in the formation of the monarchies and nation-states of Europe would seem to have been performed by other religions in other times and in different parts of the world. If we say that the Church in the middle ages guarded the poor by prohibiting usury, we must remember that in the *Arabian Nights* the usurer is equally hated, and the Jew occupies a position that is curiously parallel with the one which he held in western Europe. The Old Testament, ancient Greece, and hints from other parts of the world would confirm the impression that the condemnation of usury is the natural characteristic of an agrarian society, religion (whether Christian or not) tending to give a legitimate support to the social ethic that the times required. If Christianity acted as a conductor between the Graeco-Roman civilization and the barbarian world that supervened, it would seem that Mohammedanism served in a similar capacity for the vast dominions of the Arabs. When religion serves as one of the bonds that hold a people together it tends to operate in these various ways, apparently, even when it is not the Christian religion.

There are dangers, however, when religion is employed as the cement of society or the bond of the tribe. And we who condemn the pagans of the twentieth century for their contempt of human personality may forget that Christians themselves have sometimes found means of avoiding the implications of their own teaching. Protestants and Catholics of the sixteenth century provided precedents for an attitude which regards the enemy as sub-human; for if the enemy was a heretic was he not an abomination to be destroyed or a devil in human mask? In these circumstances the very word religion possessed certain implications which would be too illiberal for presentation to the twentieth century.

The truth is that we must consider the domination which the Church so long exercised in Europe as a thing not unrelated to the stage of development that had been reached by society and civilization. Once this has been recognized then we can say that it was the good fortune of those peoples to be under tutelage and that the greatest good fortune of all was to have as the presiding genius the Christian Church. In other words the particular ordering of things which we associate with the middle ages would be difficult to imagine save in an interim stage in the history of a civilization, and is valid and justifiable by virtue of the situation that existed at the time. It was fortunate, further, that particular circumstances in western Europe gave the Church an unusually independent position in relation to the secular authorities—an advantage it has not always equally enjoyed in modern times. We may even go so far as to hold with Professor Foakes-Jackson that the power of the Popes in the middle ages lay in the dispensation of Providence, and yet that the breakdown of that power in many parts of Europe in the sixteenth century was equally in the purposes of God. All the same, just as the conversion of the Emperor Constantine had the disadvantage that it was calculated to induce every court-sycophant to make a profession of Christianity, so

the idea of a Christian Society was bound to have its seamy underside in the rough-and-tumble of the world.

In any case there seems to be something in the processes of
civilization which leads to a greater differentiation of personality,
greater autonomy in the individual, a multiplied range of functions for men in society, and a higher claim (a claim that will be
asserted by wider classes of people who had been cowed and
submissive in various respects) to make one's own choice of a
philosophy or a faith or a view of life. At this stage in the story
habit and intellectual indolence and the group-spirit are less able
to guarantee the unquestioning acceptance of an inherited creed.
The unanimity in favour of a traditional system is broken, and in
the new condition of things the very word religion loses a certain
fringe of meaning which had come to be attached to it. In particular religion tends to lose part of that quality or that influence
which had made it so serviceable when it had had the additional
function of being the bond of the tribe.

It is not for Christians, however, to regret this transition in
itself, especially as it is one which in a sense brings society and
civilization to a finer blossoming. Their faith is one which pre-
sumes this lofty view of human personality. Indeed, for Christianity this whole transition is rather a return to the original
purity of its first state—it is a reduction of religion to its essential
meaning where it is not regarded as useful by reason of the support which it gives to something else. Christianity may be two
thousand years old but it did not arise in a primitive world or in
the form of anything like a tribal religion. It emerged when civilization was in a stage of high development, emerged as a victory
over all forms of tribalism, and addressed the adult judgment of
men not regarded as under the tutelage of custom or of group-
opinion. It did not ignore the question of human solidarity, but
solved it for a world of freer personalities at a level higher than
the herd-level—seeking to achieve the fusion by a voluntary love
which far from submerging the individual carried personality to

a higher power. We may say on the one hand that there is nothing to prevent an individual from achieving the greatest spiritual heights and giving the finest Christian witness in any century of the Church's history—nothing to prevent any individual Christian from showing the highest respect for the personality of others. On the other hand, if we are thinking of the ideas which pervade a whole society and form part of the environment of men at large, we may say that it requires a high degree of advancement in a civilization to achieve anything like a general social recognition of the value which Christianity places upon personality.

Since human beings are so wilful it may be true that the modern western world, by giving so much rein to individuals, is a civilization perpetually in jeopardy through an excess of liberty. It is a question whether this emphasis on human personality is a feasible thing in fact unless it is accompanied by a powerful affirmation of the spiritual side of life. There is grave danger for our world if, in the new situation, individuals do not by an autonomous act of judgment go over to that Christian religion which their forefathers so long accepted perhaps by habit, contagion and submissiveness. Christianity itself indeed offers a higher challenge to every individual and makes a higher test in the new situation of the world; though the response when it comes has sometimes an added authenticity, such as would come in the ancient days when a merely habitual Christian saw the light with peculiar vividness, and was converted.

If we ask what part Christianity has played in the general development of our civilization, there is a moment of transition which gives us a glimpse of the processes at work, and confirms us in the view that we must look beyond the direct policies and achievements of ecclesiastical systems, and think rather of the leavening influence of subtly pervasive ideas. A great progress in what we might call the amenities and urbanities of our civilization seems to have taken place in the eighteenth century; and roughly speaking in that period we see the effective emergence of many of the characteristic features of that "western way of life" which we so often contrast with the system of Soviet Russia. Especially we see a growing consciousness that all men—even classes long oppressed—should be conceded a larger realm for the exercise of moral decision and personal choice. These ideas emerged very often in opposition to ecclesiastical prejudice or ecclesiastical authority; and yet constantly, in the very arguments that are used, we see in them a kind of thinking that had been seriously affected by Christianity. Men who might be indignant against the Church and even deserters from the faith repeatedly betrayed the fact that Christianity still governed their unconscious assumptions and still lingered on a sort of after-music in their minds.

In the light of more sinister and radical intellectual movements that have supervened we can more readily recognize to-day how much of the thought of the eighteenth century showed signs of being an unconscious attempt to translate parts of the Christian tradition into secularized terms. Modern movements of liberalism, humanitarianism, internationalism—modern cries for freedom of conscience and for the uplifting of depressed classes—have borne marks of the leavening influence of Christian thought and it was their tragedy that, like modern humanism, they thought that they could live of themselves, and broke with the religion that had been their mother. Even in the realm of what we might call purely worldly-wisdom they cut themselves off from other parts of the Christian tradition which—as we to-day can more easily recognize —it would have been at least prudent for them to remember. It was unfortunate for the world—unfortunate particularly for France—that though there was a need for a brake on the eighteenth-century reformers, ecclesiastical opposition in particular also had its wilful aspects, and ecclesiastical authority had hardened its heart. The result was that the European tradition henceforward became too sharply divided, though not so sharply in

England, where, for example, nonconformity afforded something of a bridge between the sundered parts.

It is one of the paradoxes of modern history that the primary demand which Christianity must make of any social order, the demand for freedom of conscience, was a demand which came in the western world with every sign of the leavening of Christian influence. Yet the demand was not only resisted by ecclesiastical authority; it had to be asserted by rebellion against the older ecclesiastical organization of society. Indeed it could only be achieved by the actual overthrow of such an order. No better example could be given of the way in which Christianity may be conspiring with the development of civilization to achieve a good, even though Christians themselves, or ecclesiastical authorities, are making wrong decisions on that very matter. This means, furthermore, that those who merely preached the Gospel without *arrière-pensée*—preached purely for the salvation of souls—always worked better in the cause of civilization than they ever knew or intended; just as those who spread piety, humility, and charity are acting like a leaven, and helping to produce a different conception of personality from that hard and arrogant one which seems to be emerging in Soviet Russia. Perhaps only they can prevent the western regime of modern individualism from rushing wilfully to its doom.

It may prove to be the case that judgment has fallen on that regime of liberty and democracy which the west so stoutly defends against the modern barbarism of the east. In that case the clock is put back and the expansion of the area of this modern barbarism may bring upon us a further instalment of the Dark Ages. In such a retreat of civilization it is not possible to see how, humanly speaking, the Church, as one of the institutions in the world, may be expected to play this time the remarkable part which it was performing a thousand years ago, when the downfall of the Roman Empire in the west left ecclesiastical authority in a particularly strong position. The barbarian invasions of those early centuries not only magnified the secular leadership that was open to the Church, but provided populations more amenable to religious guidance than the barbarians of the modern world are likely to be. We shall not expect that once again for the fight against illiteracy, the provision of competent administration, the development of schools and universities, the cultivation of the arts, the promotion of legal studies and political theory, the world will be dependent on the services of the clergy. And if the urban developments in society seem to have carried with them in European history a tendency to secularization—seem always to have encouraged the rise of a lay spirit—the men of the new Dark Ages will be a still more difficult problem; for it would have to be an almost unimaginable catastrophe that could now prevent our civilization from being an affair of towns and cities.

If the new Dark Ages come Christianity still carries forward—as it did before—the legacy of the higher civilization. It does so not by confronting communism with an authoritarian counter-system, but because it cannot help asserting the uniqueness and the value of personality. Precisely by pursuing its task in the salvation of souls, confident that in this case once again Providence has unexpected things in store—precisely by seeking the kingdom of God and waiting for the rest to be added unto it—the Church can repeat the glory, though it can never expect by sheer contriving to duplicate the actual achievement of a thousand years ago.

III

Christianity in the
Twentieth Century

14

The Challenge of the Faith

It was never my intention to set up as a teacher in religion, and I always have the feeling that heaven will strike me dead if I show presumption in this respect. On the other hand it is arguable that the Christian ought not to be unwilling to make confession of what he believes; so that I am between the upper and the nether millstone, for I also have the feeling that heaven will strike me dead if by repeated refusals I seem to be declining to testify to the faith. Let me say that I do not pretend to speak with authority and I am not to be regarded as an authority. And, speaking as a layman, I see no reason why particular views that I hold should carry conviction for anybody but myself. On the whole I shall try to treat my problems from the point of view of the position which I, as an historian, think that Christianity occupies in the world at the present day.

Granted the present stage of civilisation and the present state of the world, I would say in the first place that we should be standing on very doubtful ground if we expected men to believe in Christianity because their fathers wanted them to believe in it. There is some point in a man's playing cricket to please his father, and it is allowable for a man to become a lawyer because the family tradition is a legal one. But the imposition of the ideas of one generation upon another through the tyranny of the older or by a narrow process of in-breeding or even by some *esprit de corps* in a family is liable to great abuses and belongs to a more primitive state of the world; and I doubt whether the principle is going to be helpful to religion in these more modern times, when it might just as well entail being anti-Christian because your father was anti-Christian. It may seem very well to have a section of our globe which through the power of mere tradition stands from one generation to another in obedience to the Church. But if that is the rule you go by there will be a further section similarly submerged under a sort of hereditary Mohammedanism; and, after all, in the days of the pagan Roman Empire, Christianity could hardly have propagated itself except by calling on men to forsake the faith of their fathers. I do not think that I was likely to do any great harm in the world by becoming a teetotaller because my father wanted me to do so; but it would almost be sacrilege to pretend to believe in Christianity in order to please a parent—to pretend when one was not really convinced of its truth. It would tend to pack the Church with unconvinced and halfhearted semi-Christians, who are a crushing load for any Church to have to carry and are calculated to be a menace to the whole cause. The most that Christian parents can legitimately require of us is that we shall not leave the claims of the Christian faith unexamined, or lapse from religion through mere back sliding, or catch paganism from the world by mere contagion, or betray our heritage through mere levity and indifference. If ever anything assumed and demanded the voluntary act of the individual it is Christianity—a religion which addresses itself to what is most sovereign in human personality.

In the second place, it is necessary now to recognise the fact that it is not possible to accept Christianity on the ground that so many people believed in it in the past, or on the ground that it prevailed for so great a number of centuries in our part of the world. For some people this must mean accepting Christianity merely in the way that one continues a custom. This in turn has had harmful results for religion itself, for it tends to produce merely conventional Christians in a conventional world, and such people tend to accept anything else that happens to be customary, so that religion comes to be tied to all sorts of conventionalities, and intelligent people naturally come to see Christianity itself as another tedious and artificial survival from the past. In any case, those who in one age accept Christianity out of mere custom or indolence of mind or submissiveness to existing conventions or inability to think things out for themselves will be equally liable to become unbelievers by mere habit and routine when the prevailing tendencies in the world are pagan. We cannot rest the claim of religion on mere numbers and weight of opinion, cannot say that the bulk of mankind in the past or the present have been Christian. And even if we confine our attention to the European section of the globe, such arguments have ceased to be convincing. After the downfall of the Roman Empire a great deal of Europe was Christianised either by the power of the sword or through the command of princes or under pressure and persecution. For centuries this part of the world was kept Christian partly because in a more primitive state of society men more easily accept what their fathers have handed down to them. Most of them were not really allowed to know that there was any alternative to Christianity, and indeed, if there was any alternative, it was only a relapse into the darkness of pagan superstition.

Perhaps it is more common for people to say that they believe in Christianity on the authority of the Bible or on the authority of the Church. This is an argument which always gives me a certain amount of bewilderment, since it can so easily be inverted. After all, one might just as easily say that it is only because one believes in Christianity that one accepts the Bible or the Church at all. The argument is all the more confusing in that Christians themselves may be said to be divided into two great historical sections—the one says that we only believe in the Bible at the command of the Church, that is to say, because the Church established its authority; the other says that we only believe in the Church because it is grounded in the Scriptures. Some people seem to combine and confuse the two arguments—if you ask them why they believe in the Church, they say that it has scriptural authority; and then when you ask them why they believe in the Scriptures, they say it is on the authority of the Church. But whether the tortoise is standing on the elephant or the elephant is standing on the tortoise, one must ask whether either of them is standing on firm ground. A man who was not brought up in unquestioning belief from the start, or a man confronted by the claims of different religions will want to know why to believe the Bible rather than the Koran. In one part of the world men say: "This is the Word of God"; in another part of the world men say: "No, it is we who have the word of God." Let us make quite clear at the start that when a human being comes to us and says "God demands this," "God insists on that," such a human being is to be submitted to more careful and severe criticism than any other preacher or hawker in the world—for, otherwise, this poor human race is going to be bedevilled by false prophets. On the other hand, we must be equally careful not to make gods out of sticks and stones or mere abstract nouns, for those who refuse to believe in a personal God tend to deify something mundane, something which we can easily discover to be far from divine. Even the twentieth century can have its blind superstitions.

The present age is a hazardous one, but also an exhilarating one, because at any rate now we cannot go on evading the fundamental questions. And all Christians anxious for the spread of the Gospel should be eager to drop all hedging, all merely comfortable formulas.

The whole problem is more difficult for modern man in that, whether we find it in the Bible or in the Church, the word of

God only comes to us through human agencies, and nobody can deny the imperfection of these. There seem to be contradictions in the Bible and irreconcilable pronouncements; there are signs of primitive states of society, of ideas that only gradually developed, of a spirit which we today find it difficult to defend. And something of the same is true at the same time of the Church itself—any claim to divine authority on its part seems to me to be not instantly plausible to the ordinary reader of ordinary history. In the Church there has been intolerance, persecution, superstition, attachment to convention, a spirit of worldliness, and a terrible lack of charity at times. When great and obvious mistakes have been made people may say that it was not the Church which was mistaken—merely a bishop, a human agent, was wrong in thinking that he was speaking with the voice of the Church. I have heard Marxists argue similarly, when crime and error have been committed, that it was not Marxism, not their Church, that was wrong, but only the human agents of it that had been proved defective. On that argument a Bishop today, as in the sixteenth century, may say that such and such a thing is the voice of the Church; but here again, if in future ages there is a modification of the view, it will still always be easy to say that the Church was never wrong—the Bishop was merely mistaken in his view of what the Church had said. All this is not very nourishing food for a man who is asking himself why he should believe in the Church at all.

In the modern world, therefore, one cannot take short cuts with ultimate questions and nothing can save us from the necessity of confronting the naked issue. In a most definite and intimate sense the issue is between a man and his Maker—no man can be a Christian on the strength of somebody else's belief in the faith. All of us have to confront the question: What God will I worship in life and what moral end am I going to serve and what faith am I able to discover, what faith to live by while I am in the world. And we in the twentieth century are especially handi-

capped when we face these fundamental issues. We are surrounded by such a multitude of distractions that so often we
seem not to have time to spend on the deeper questions of life.
In any case we are not trained in the deeper kind of contemplation. The present-day world has lost the sense of the value of
contemplation in itself; and we are beginning to lose those deeper
truths that only come from contemplation, from a rich internal
life. In this sense, the people who lived at the edge of the desert
would seem to have examined themselves more intently and
pondered more deeply on the mystery of the stars, so that they
may have something to teach us, something which complements
our own kind of experience. Those who lived when the world
was static—when the silhouette of the ploughman against the
horizon hardly changed in the passage of a thousand years—may
have something to teach us, who only know a breathless, rapidly-
changing world and who seem to be having to pluck what we
can from life while running at full speed. One of the reasons
why I think that history is valuable is that it helps us to grasp
something of the totally different experiences of life which other
men and other ages have had, and which complements our own.
Furthermore, we should bear in mind that though, since the
Scientific Revolution of the seventeenth century, we have come
to put our passion, our industry, and our ingenuity into the study
of inanimate things—or into the study of human beings as though
they were things—there have been ages of history when men
were less concerned with things, and when indeed they had fewer
things to concern themselves about. And if, for one thing, they
lived nearer to nature, they also put more reflection and ingenuity
into the analysis of their own internal life, into the study of personality, and into the search for wisdom on the subject of human
relations. I am not saying that we must have the humanities instead of the sciences, because I firmly believe that we ought to
have both. I am not saying that we ought to be superstitious
about the wisdom of the past or the authority of the ancients. I

mean rather that we ought not to be the mere frothy children of the present-day, but ought to try to collect something of the experience of all mankind. We ought not to be content to be merely the child of twentieth-century fashion when it is rather our function to stand as the heir of all the ages.

In this connection let me say that if ours is the age of science and technology—if for a certain space of time the human race is going to concentrate on that side of life, and indeed has no choice but to concentrate on that side—there was a period from about 600 B.C. to 300 A.D. which I suppose is the greatest age in the religious history of the world, the greatest age not merely for us who inherited Christianity but for wide ranges of civilised peoples in Asia. I see no reason why one should not begin by looking at this subject in quite a mundane manner—granting that men in those ancient days were prone to superstitions about the world and nature which the modern scientist has done so much to dispel; granting also that whatever deep and permanent truths they discovered were entangled in the superstitions and the notions of the time; but asking whether at the same time these people did not reach truths of such profundity and such enduring momentousness that the world has rightly clung to them ever since. Believers in such truths are not to be regarded as mere blind idolaters of the past.

In those ancient days men, who on any interpretation were certainly not less mighty in intellect than any of our modern geniuses, set their minds to the consideration of human destiny and of the inner self of human beings. Rising above the materialistic superstitions of a pagan world they asserted the truth which more than any other serves to disengage religion from primitive crudities and sinister entanglements—the truth namely that there is a higher grade of existence for human beings in the spiritual realm and in communion with a God who is Spirit. Just as some people, living from hand to mouth at a low level of existence may never suspect that there is a higher life in the realm of the intellect and may even jeer at the people who prize intellectual pursuits, just as some may be intent on what they eat and drink without ever knowing the profundities of poetry, or may devote all their efforts to making money and never have an inkling of the magic that lies in music, so men (even those who are intellectualist) may by their blindness be ignoring a more lofty state of existence still, closing their eyes to a further dimension that is open to human life in the realm of the spirit and in contact with spiritual forces. If anybody is sceptical about this, let me say that I do not believe that people seriously deny the extraordinary power that has been manifested in spiritual men in history. What the world questions is not their power, and not the amazing height to which they carried human personality, but their way of explaining themselves. The world questions their self-diagnosis when they tell us that they owe their strength and their internal resources to the spiritual realm and to their contact with God.

Now I think that nobody can have taken more pleasure than I in that intellectual revolution of the seventeenth century which transformed the study of the external universe and which constitutes the origin of modern science. One of the basic principles of that revolution was a certain purification of scientific procedure—the policy of concentrating on the process of nature and the physical world, on the behaviour of certain bodies when heated, for example. They said, stop confusing the issue by mixing into the discussion the theories that one might have about the meaning of life or the final purpose of the creation. By concentrating on measurable material processes it proved possible to attain such stupendous results that men have devoted more and more of their intellectual activity to the examination of the physical universe ever since. But the men who produced that intellectual revolution did not by any means intend to assert that only the physical universe exists. On the contrary they were mostly sincere Christians, proclaiming all the time that they were working to the glory of God. It was the succeeding ages who became the slaves of their own techniques and the victim of mere habits of mind, so that they imagined that nothing could exist save the things which their instruments could weigh and measure. It is as though the scientists were to discover a pipe organ and examine all physical parts of it and count the vibrations it could produce, and measure the thing this way and that—until the time comes when nobody can remember for a single moment that here was a thing which could move human souls by emitting the music of Bach.

Both history and our own childhood will show us that man does not begin his thinking with a stark notion of a purely physical universe composed by the cooling-off of incandescent gases. That is a specialised notion of the universe, produced by a process of intellectual abstraction—produced by adopting the excellent scientific principle that for the purposes of physical science you will study only the physical universe. If you begin with the assumption of a purely materialistic universe it is true that it is very difficult to tack a God on to that, but also it is very difficult to tack morality on to it—and that is why morality tends to weaken its hold in a materialistic world, for morality then becomes merely a man-made thing, and easily slides into being a pliant tool of the state. If you begin with a universe composed of asphalt and volcanic dust and boiling gases, of course it is difficult to bring God in afterwards, or spiritual things as an appendix or morality as an extra. It is like trying to plant roses in a bed of concrete. It is possible therefore to have caught by mere contagion—that is to say, without ever thinking the matter out—a view of life which has short-circuited the problem of the existence of God and has virtually eliminated that possibility in advance. We live in a universe which seems to follow laws or seems capable of reduction to law; and I do not know of any alternative to the view that such a universe either comes from God or is the product of chance. Either it gives us a hint of the mind of God or else we must just say that particles of matter, in an eternity of permutations and combinations, have happened to hit on what for the time being seems to be an unaccountably orderly system of interlocking. About all that, I only want to say: Be careful that we are not begging the question from the start.

The Bible picks up the other end of the stick, and from fairly crude beginnings, it develops a theme which in the New Testament acquires the might and majesty of an orchestral symphony. The subject is that of the spiritual life disengaging itself more and more from the materialistic superstitions of ancient polytheism. The whole theme is a triple one, concerning Love, concerning Personality, and concerning God; and all three grow together, men forever learning more of any one of them from what they learn about the other two. The deepening knowledge of God makes one's appreciation of human personality more profound and adds to the richness of the idea of Love; but all the time it seems that all the three are reacting equally on one another. The Love, as it develops in the Bible, is not merely what we call natural affection, for natural affection—even the love of parents for children or of man for a maid—may become warped and disproportioned and bitter. God and human personality emerge as spiritual, and the Love itself is the loftiness of spiritual Love, the higher regulative Love of the New Testament, itself a theological virtue and a spiritual force. It is a thing which, as seen in some of the saints, should be regarded as one of the evidences for a spiritual view of personality. If Love and Personality are values to us, as many people say, then it is these that our minds ought to seize upon when we are seeking out the nature of the created universe—these and not the mere clay and the incandescent gases. And though belief in God is a matter of faith, a case of choosing the faith to live by, still I think that these three—Love, Personality, and God—do belong together. They are ideas that grow together in us, and each of them in turn seems to wither if men neglect the spiritual side of life. It is at least an easier thing to make the transition to the idea of God if

one feels the universe as a warm thing, throbbing with personality, and not merely a wide waste of molecules that seem to be
equally at the mercy of chance and blind necessity. And the picture that is produced in the Bible is one of God presiding over
this world of tumult and violence, of cupidity and fear, of
struggle and cross-purposes—presiding over it and drawing upon
it like a magnet, pulling at wilful men with the cords of Love.

I do not believe that it is science as such which has undermined
faith or that those who have deserted Christianity have done so
through meditating on biology or astronomical systems and pursuing scientific principles to their limits. The conclusion that we
come to depends much more on things that are happening inside
ourselves and things that we see when we look inside ourselves.
The real problem is the issue of man's internal life. I think that
the greatest problems of life and the universe can only be solved
if conditions inside ourselves are right. Nobody believes in the
intellect more deeply than I do, and I insist that when we consider life and the universe we must start from this—start from
personalities conscious of themselves, lumps of clay that actually
ask questions about why they are here. But even in the non-
religious field many people would agree with me that over great
ranges of subject-matter one will never reach the right answer
without intellectual humility; there is an intellectual pride which
itself produces aberration. If we forget that ours is only a limited
intellect, capable only of broken lights and partial visions—if we
think that our intellect is the King of the Universe—we certainly
will not get the right answer to any question relating to God.
Also if our mind is in rebellion—rebellion against our parents,
against society, against the universe—we will not get the answer
right. There are many fields, some of them in the realm of science
itself, where it is only the affectionate mind and the sensitiveness
of sympathetic imagination which can ever put us on the right
track at all. Also there is some point in having reverence for the
great profundities—in standing breathless for a moment before

the beauties of nature or the music of a starry sky—in realising that we live every day of our lives next-door to a great mystery. We have not plumbed the depths of this universe in the three centuries of modern science. And, in spite of the painful dogmatisms of some religious people who seem to think that they have the answer to everything, I think it is religion which pays due respect to the unfathomable mysteries.

All this, however, as I have said, is merely a case of setting the conditions, for thought about this matter and belief in God is like belief in the universe—in fact, it is belief in the universe and indeed belief in personality, belief that we are not mere combinations of particles, not the mere product and sport of chance. The real challenge of Christianity comes elsewhere; and we find the evidence of it throughout the Bible, though in this as in everything else the climax comes in Jesus Christ. In one of its aspects the Bible is a great developing treatise on the problem of sin and righteousness; and it is remarkable in many respects, for I do not think I know of any people which in the national history that it has handed down from generation to generation, has made it the story and the confession of their own sins in the way that the Old Testament has done. I do not know any point on which the modern mind is more superficial than in respect of national morality or political morality, any point where we are so careless in the education of the young or the enlightening of public opinion. Now in the great period of religious history from 600 B.C. three important things occurred. First of all, whereas the nation as a corporate body had had its God and had been arrogant in its exclusiveness and self-righteousness—expecting God to bring it victory and material prosperity—you had the transition to the view that God had contact with each separate individual, that religion was a matter for the inner man, that the law was written in the hearts of human beings, that worship was a spiritual thing—not a matter of bribing God to give you material prosperity. Secondly, righteousness lost its arrogance and came to be seen as

a matter rather of affectionateness and self-sacrifice. It was discovered that there was a higher morality than that of mundane
society—that in the light of this all men are sinners and even all
our righteousness is as filthy rags. The extraordinary thing was
that, precisely because everybody now had to acknowledge their
guilt, men became kinder to human sin, regarding it more as an
offence against God, not a thing that called for vengeance and
violence on the part of men. In one sense everybody had to feel
guilty, but in another sense you did not feel paralyzed by guilt
or under a curse. And more and more the talk was of forgiveness
and mercy; the highest poetry, the highest ideal was when you
talked about the good man taking upon himself the sins of others
—all this long before Christ. Thirdly, instead of a God judging
with fire and thunder you talked of a God drawing men with the
cords of love.

It is this higher view of righteousness and this more merciful
attitude to sin itself which is the least studied at the present day.
Above all it is when all this comes to its climax in Christ himself
that it becomes a tremendous challenge to heart-searching in all
of us, since, in a sense, it is the basic theme of the New Testament. Here is the picture which, when taken to heart, touches
the problem of moral progress at the key point, by convicting
men of their own sins and convincing them that here is the most
important of all the factors that spoil human destiny, and at the
same time revealing that there is a higher order of life, a higher
grade of spiritual being. Here are the things which make one see
the reality of spiritual forces and make us feel that God is tugging
at us with the bonds of love. I conceive myself in an essay of this
kind to be talking principally to people who are Christians already, and to them I would say that I believe that here, at the
problem of sin and righteousness, is the strategic place at which
the world is capable of being touched. Even those who have not
puzzled their way through modern science to a doctrinaire belief
in God know that they have a conscience; they may not be able

to judge between astronomical theories but they are capable of passing judgment on themselves. The first step in the Christian life is not a theoretical one at all; it is just a realisation of one's sinfulness, a realisation that possibly human sin is the thing which hides God from us and estranges us from Him. In other words, the authority which the Bible has for us in so many of its parts is not a matter of superstition or authoritarianism but a sort of self-ratification that it carries—a feeling that this teaching speaks to me, cutting deeply into my deepest experience of myself and expressing my highest aspirations. Even if the New Testament were a work of fiction and were presented to us as the achievement of an imaginative poet, its disturbing challenge and its exposure of the human plight would still be as valid as ever they were, and as vivid as any Shakespeare could make them. I believe it is true that this must be the ultimate basis for all, whether they ultimately become Catholic or Protestant. We may have to change much of the barbarian language and ideology which it has been customary to use, and of which I see no signs in the Gospel; but a Church can never get to grips with the outside world unless it dares to attack the whole problem of human sinfulness, not after the manner of our mundane propagandists, but in the manner of Christ.

15

The Obstruction to Belief

I would hold to the view that the time has come when Christianity must compete on fairly equal terms with all the other gospels, creeds, and ideologies which now do battle for the possession of the souls of men. The Christian Church has been driven from that general presidency which for special reasons it enjoyed for so many centuries in European society. It has lost that majority position by means of which it had acquired so many monopolies and priorities and special advantages for itself. Against rival programmes and hostile intellectual systems it now has to make out the kind of case that will carry conviction to the outsider. And when it presents its case it must confront the fundamental issues; it can no longer reckon on a bias or a predisposition or a prejudice in its favour on the part of society as a whole.

The situation, however, is much more difficult for the Church than even this description of it is likely to suggest. I personally would strongly hold to the view that men are not naturally Christian and that at the best of times a strong resistance will be made to the essentials of the faith, a resistance which can be associated with ordinary worldly-mindedness. At the present day we must say that, over and above this regular reluctance, Christianity has to meet a special obstruction for which our own predecessors are responsible and which springs from the defects in the Church's history. Christians themselves, it is true, are able to see how such defects could have come into existence, and they are in a position to avoid being unduly resentful about them. But one might say that it takes a Christian to understand the errors and the evils, or to make allowances for the men who were responsible for them. In the case of the people outside the Church these things have had the effect of producing such a blockage of sympathy as amounts to an obstruction to the hearing of the Christian message. And men who are disposed to be deeply moved by the figure of Christ in the Gospels, will decline to go any further, decline really to come to grips with his teaching, because they think that by becoming Christians they would be reviving or reinforcing a sinister vested interest.

There are two ways of looking at the Christian Church in history, and they seem to lead to results which are antithetical and irreconcilable. In reality they represent two different levels of analysis; they are the effect of cross-sections taken at widely different points. One of them will appear amongst Christian polemical writers; the other belongs rather, I suppose, to the non-Christian who calls himself a Humanist. In my view it is necessary for us to adopt the two lines of approach and to follow them concurrently; because it is when the two are properly taken into account and their results are brought together that interesting things are likely to occur.

On the one hand, I must confess that the long story of Christianity seems to me to be in a particular sense the continuing revelation of the nature of God. In this sense, it is a prolongation of the Bible, and the texture of European history is in continuity with that which we see in the narratives of ancient Hebrew times. I do not believe that God showed himself in human history in the days of the Old Testament and then, from the first century A.D., began to withdraw from the world. Nor do I think that

Old Testament history differs from modern secular history except in the sense that the ancient Hebrews were able to see the hand of God in the story and were determined to put a religious interpretation on the events that took place. Some people seem to think it a special tragedy that Jerusalem, the Holy City, which one would like to imagine as the haven of peace, should have become the scene of so much violence and tumult in recent decades. I personally think that it is truer to say that Jerusalem has become more like its old self again, more like what it was in the days of the Old Testament. For Jerusalem even in its greatest days was a very violent city; the prophets said that it was a very wicked one to boot. For this reason it came to terrible destruction more than once; and if it was ever quiet it was only while it was under its thousand years' enslavement to the Arabs and the modern Turks. Yet it was precisely under such conditions of turmoil that men learned to discern the hand of God in history.

Taking the religious point of view, I should feel that in nearly two thousand years of European history, the existence of New Testament Love, the spectacle of Christian humility, and the knowledge of forgiveness of sins have been like something supernatural superimposed upon the mundane story. Indeed they *are* an element of the supernatural added to the human drama and intermixed with it, with results that are evident in the lives of saints throughout the ages. Even a purely secular analysis of the story has to give recognition to the power of these things—however it may interpret them—and I think such an analysis serves to show how it is just these that come to be vindicated over long periods of history. It was a mere handful of men—generally with only limited mundane qualifications and coming from a people regarded as odd and unattractive—that set out to convert the ancient Roman Empire to Christianity. It was a remarkable spirit, and one capable of generating very original things which issued in the Gothic cathedral of the middle ages or in the poetry of Dante's *Paradise*. The deepest principles of our religion have worked like a leaven in European society, and we can trace with a microscope the way they helped to build up what we call our Western "values"—the values to which even the secular world is attached. Many of those who reject Christianity today do in fact hold a quite different ideal of human nature from that of the ancient pagans—they have a different way of mounting the whole human drama and conceiving the rôle of man on the earth —for the simple reason that fifteen hundred years of religious predominance have permanently altered our angle of approach. Sometimes Christian charity has been operating for good in the world, even while Christian leaders themselves have been wilful and have been working against that good. And throughout the centuries the light of the Gospel has never been extinguished and never lost its relevance for the world. Wherever a man has sought to grow in piety, the Church, however bad its condition, has always been able to provide him with the means.

The other side of the story is a dark one, however, and it is natural that only those who are Christians themselves can understand it or make it square with their view of the way in which things happen on the earth. I sometimes wonder whether the Christian revelation can have been in the world for half a day without somebody seeking to discover either how to make money out of it or how to use it in order to acquire power over his fellow-men. The organised Church, though it had had to claim liberty of conscience for itself when propagating the faith in the pagan Roman Empire (and though voluntariness was the very essence of the New Testament religion) insisted on persecution from the very moment the policy became feasible, and then, after fifteen hundred years of authoritarianism, it fought for the continuance of every single weapon of persecution, one by one— fought even against the secular state and worldly minded men who wanted to stop the practice because it was too cruel and barbarous. Even the Communists have shown greater restraint— even the Communists would have to go much further than they

have done in their measures against Christians before reaching the
degree of cruelty which they themselves would have had to
suffer so long as the Church remained supreme. Our great Cambridge historian, Lord Acton, goes so far as to say in his manuscript notes that though he discerns the hand of Providence in
secular history and in the general course of things, he cannot see
it in the actual history of the Church. I do not agree with him,
but I understand what he means; for if we had been living in the
days of the Twelve Apostles and had been asked to imagine all
the perversions and abuses to which Christianity might be liable
through its entanglement in the world, a present-day survey of
European history would show that all the predictable evils had
actually taken place, but also more still—indeed the devil turns
out to be more ingenious than anybody can ever conjecture in
advance. Providence did not stave off the evils; but where Acton
was wrong, in my view, was in his failure to make due allowance
for the achievements of the saints and the triumphs of the spirit,
which are so remarkable in spite of the inevitable imperfections of
any organisation that is entangled with the world.

I have sometimes thought that a religion claiming exclusiveness
is a terrible infliction on the world unless it is of the essence of
that religion to give charity the presiding place in the conduct
of mundane affairs—a presiding place which we can hardly say
that it enjoyed in the actualities of our Church history during a
great part of the story. I think that the persecution of the Jews
is perhaps the most contemptible and pointless of all crimes; but
I can hardly avoid asking how far its existence in our civilisation
is not connected with the long predominance of an intolerant
form of Christianity. Such a form of religion can hold back both
the amelioration of society and the advance of civilisation. Sometimes religious inhibitions, or too hard a literalism in the use of
the Old Testament may have prolonged a piece of barbarity or
stimulated acts of cruelty. I am shocked to find in history that
on occasion worldly minded men have asserted a terrestrial mo-

rality against an alleged supra-terrestrial morality; and we at the present day are glad that they succeeded—even Christians are glad—and sometimes we count it today as a success achieved on behalf of Christianity itself. Ecclesiastical interests have often operated against the amelioration of the conditions of the lowly, especially when the Church has become tied to the support of a nation-state or a particular social régime. When charity is deficient the idea of Providence itself may be used to keep the humbler classes down and to support vested interests in their resistance to social improvements. Ecclesiastical leadership has often been unwise—resisting such things as freedom of conscience, democracy, and egalitaranism, though nowadays we tend to speak of these as Christian ideals. And because the Church has been on such a wrong basis in its relationship with the world it sometimes cuts a pathetic figure in modern history—resisting what is new, opposing progress, frightened of science, conducting a long pathetic rearguard action, only to be beaten in one century after another.

It should not be impossible for us, therefore, to understand the attitude and appreciate the fears of the non-Christian Humanists, who tell us that they dread the return of obscurantism, and who at least call our attention to the paradoxes of Church history.

When Christianity has prevailed as an intolerant exclusive religion, there have been cases, I think, where it has been entangled in principles calculated to produce cruel action on the part of men not naturally cruel—so that something in the system as well as in the agents may have been at fault. When all things are considered, I must confess that, if the occasion could be plausibly regarded as imminent, I myself would share some of the fears that the non-Christian Humanists have so often expressed. I, too, would dread the possibility of a return to that form of society in which the ecclesiastical profession had the predominance. And though Christians today are often so very different from their predecessors, I am not sure that they realise how profound the problem is—I am not quite convinced that they have renounced as much of their history as would be necessary to ensure the removal of the danger. I have such admiration and such a special affection for my friends among the clergy of all denominations that I would not want them for a moment to imagine that I impute to them secret dreams of power, or indeed any trace of mundane ambition. But when we are taking long-term views we have to remember that people who are in one kind of situation never seem able to imagine what they would be wanting to do if they were in a different situation altogether. The Methodists say that they have never persecuted and I am prepared to believe that they have never wanted to persecute; but they came into existence after the date at which persecution could have been a thinkable policy for them, and as they have held at times some of the basic principles which are capable of being developed to the point of persecution, I am not sure that I would claim for them any special virtue in this connection. Perhaps we must rather say that they were fortunate in that they did not emerge until the eighteenth century. The Humanists do not feel assured that the Christians would not abuse their authority again, if ever they were to recover it—do not feel sure that Christians have entirely ceased to covet mundane power, in the sense of power to impose by force some of the regulations they consider righteous. I personally wish that Christians would rely utterly, and rely only, on the power of the spirit; but supposing the Humanists are mistaken in their fears, supposing even they are entirely mistaken in their attitude to this whole question, we must at least take their view as a measure of the deep historic resentment against the Church, a resentment so strong that it is perhaps the primary obstruction to the cause of Christianity today. I am always interested in the title of a book by Bishop Barnes, a title which reads: "Can such a faith offend?" It represents exactly the way I feel about the Christian religion. Yet I believe that one of the main reasons why the Gospel finds itself blocked—why men close their ears to it—is a deep suspiciousness and an unformulated, half-conscious resentment.

There is an important sense, however, in which we can say that the Humanists are mistaken in their criticism of historical Christianity. Their very understandable arguments have been entangled with some out-of-date Whig history, and at least they do not always make the correct inferences from the data. When I was first taught European history, no communist or Nazi régime had come into existence and we felt that we were living a secure life in what seemed to be a liberal civilisation. Indeed it might have been argued that apart from the ordinary number of criminals, human nature in general presented a fairly respectable appearance. The one intolerant and persecuting organisation that we ever heard of was the Church in the earlier centuries of European history. And certainly it seemed to us as sinister in retrospect as the communists and the Nazis were to become for a succeeding generation. Knowing nothing in those days of the totalitarianism that was to come, and knowing little of the history of the other civilisations on the surface of the globe, men found it natural to believe that Christianity alone was responsible for the introduction of persecution and obscurantism into a world which would otherwise have progressed smoothly. It is the combination of modern communism and nazism which has taught us that men will persecute and repress and will turn to obscuratism even when there is no supernatural religion—that the world will sacrifice human beings to sticks and stones or to abstract nouns as easily as to a God above the skies. The history of science has shown us that, whatever it may do in an intervening period, the Church will not go on fighting a scientific theory once that theory has really been established. On the other hand, the hostility of the Church to the Copernican theory was matched all the time by the hostility of astronomers—that is to say, by the conservatism of the scientists themselves. Even the artists and critics can be authoritarian in their oracular manner; even the scientists can be intolerant. The Humanists are right to be afraid but what they ought to be afraid of is the intolerance and the cruelty that are latent in human nature as such—the universal liability to sinfulness. Our chief controversy with them is a controversy over human nature. They lack the Christian view of the spiritual nature of man and his eternal destiny, but they rely, even against the evidence, on the goodness of men in general, and the result is that they are nonplussed when human nature shows its ugly side. What Christians themselves must learn on the other hand is that Churchmen in history, and the makers of our ecclesiastical traditions, have not been exempt from human cupidity; and the cupidity can gain tremendous leverage through the mundane power of an exclusive religion claiming authoritarian privileges. The cupidities even become incorporated into the system itself if great ecclesiastical organisations think that they are working to the glory of God when they resort to the weapons of the world. In a sense we must learn the same lesson that the Humanists have to learn; for we should find the due corrective if we could remember to regard Christianity not as the religion of the righteous but as the religion of sinners. Being a Christian does not give you the right to impose your will or your views on anybody else.

All that I have been saying has reference not to the essential doctrines of the Christian faith but to the problem of the relations between the Church and the world. And what I have been saying amounts to the thesis that in the centuries of historical Christianity the relationship between the Church and the world has been such as possibly to call for a reconsideration of the whole problem.

So far as I can see there are two rôles which religion has played in society in successive periods of history; and there is so great a difference in the way that it functions in the two cases as to justify us almost in saying that the very word religion suffers a change of meaning in the passage from the one to the other. The transition is observable even in the Old Testament where in the first place it was the Hebrew people as a group, as a corporate body, that was regarded as having a direct relationship with Jehovah. This national solidarity is the point to which I want to draw your attention; and I want to say that it is primitive in character, suitable for undeveloped societies, and common to many forms of religion, even to pagan forms; and by its nature it entails the persecution of any dissentients. But religion can develop from this state, and this would appear to be one of the reasons why the Jewish Exile is so important in the history of our part of the globe. It was the time when the people were dispersed and no longer existed as an organised political body, and, so far as I can see, it represents one of the most important movements in the history of individualism in our part of the globe. The Exile confronted men with the question whether the faith could continue at all when its seat in Jerusalem had been destroyed and the land forsaken and the people scattered in alien countries. It came to be felt, however, that God was with his children, wherever they might go; he was with them as separate individuals when they no longer existed as a consolidated group. He had condemned the nation, yet he would save the individuals who repented or remained faithful. He would judge them according to their individual righteousness and no longer punish children for the sins of their fathers. He would be their God even though the Temple no longer existed; and now it was realised that he could no longer be confined to the Temple—he was to be found in the hearts of individuals. He would even be the God of those foreign people who had never belonged to his chosen nation, and amongst whom the ancient Jews were now scattered. The thirty-first chapter of Jeremiah and the eighteenth and thirty-third chapters of the book of Ezekiel provide evidence of the development that took place. Religion, instead of envisaging the nation or the group, found its seat in the depths of the individual personality. Here—that is to say in the religious history of man, as one might expect—is to be found the roots of the doctrine of individualism. It is not the political nation but the human personality that is the Temple of God; as Lord Acton said, it is not organised society but the human being that has the eternal soul. Each man has his separate wire connecting him with eternity; each therefore has direct relations with God. It is clear, furthermore, that under this new system, religion is regarded as more definitely a matter of the inner life; God now talks of a law that is written in the hearts of men. And because religion is less a national and a political affair—less a question of squaring God in order to be able to achieve military success or material prosperity —the whole process is one in which we can see religion itself becoming less pagan, becoming a more spiritual thing. Still another point is involved in the same historical transition. For reasons which are obvious, religion, no longer tied to the corporate nation, can now begin to take on an international character.

Now the change which I have been describing must be regarded as one of the great transitions of world-history, and the point of principle which it raises presents us with one of the greatest religious issues in the world. It seems to me to be of the highest importance that both Christians and anti-Christians should be prepared to alter their traditional views on this matter. On the one hand you see religion in its pagan and primitive rôle—religion as the affair of the group, the business of the whole nation in its corporate capacity. It is religion at the service of mundane ends—religion as the cement of society and the bond of the tribe—jealousy guarded by the tribe because worldly good fortune depends upon it, and in turn it acts as the buttress of the existing régime. And of course there is no question of the individual having the right to say that he does not believe in the religion of the group—the individual is not supposed to set himself up against society. It almost seems that any religion will do for the purpose of the group—a great civilisation was built up around Mohammedanism as well as around Christianity; and possibly there were forms of ancient paganism which served more effectively than Christianity as the cement of society and the bond of the tribe. On the other hand, there is religion as a matter for the inner man, as the one thing in the world which needs an inner ratification, an authentic co-operation of the individual will—a matter for personal decision, with the individual brought face to face with God. And the solidarity which this religion requires is not a case of submission to the spirit of the herd or subjugation to an external authority; it is the solidarity simply of people loving one another and recognising that they are bound to one another by the faith they hold. All our religious history is the long story of the conflict between these two forms of the idea of religion. And we find the conflict recurring actually in ancient Hebrew history itself after that period of the Exile which I have just been describing. Even after the vision they had had of something better, the ancient Hebrews went on lapsing again into the older and more pagan type of religion, the type which was under the domination of the herd-spirit. This meant a decline of spirituality, a hardening into legalism, a fall into national exclusiveness, a corruption into belief in the kind of Messiah who would bring mundane glory to the children of Israel. Far from producing good, this decline served to hasten the fall and achieve the destruction of the Jewish nation. And Providence has not ordained that historical Christianity should be exempt from the same conflicts, the same processes of decline. What Providence has secured has been such a persistence in the more personal side of religion and such a development of its power, that the resistance to the domination of the group sprang from the first out of the religion itself, and based itself on Christian principles. Even in the highly civilised days of the ancient Roman Empire, churchmen chose to make alliance with mundane power on the first occasion that this was possible, and indeed they persecuted as soon as they were in a position to persecute. We can understand why such things happened, and we can see how plausible were the arguments used to defend the policies in question. I think that many of us might even wonder whether, if we had been living at the time, we would not have made the same mistakes. Even this, however, only makes it the more understandable that the Humanists should suspect us and should feel a shudder of terror at the thought of a revival of Christian domination.

There is another great historical event to which I believe we pay insufficient attention, when we are discussing the relations between the Church and the world; and that is the great collapse in Europe, the great hiatus in the history of civilisation, which occurred when the ancient Roman Empire fell and the Barbarians took possession of it from, say, the sixth century A.D. Though parts of Europe had been highly civilised and urbanised, the western half of the continent lost its cultural inheritance, and society reverted to primitive forms. The Church, however, which had developed under a highly-advanced civilisation, made itself the custodian of the ancient culture and set out to educate the barbarians and recover the learning of antiquity. It was not the Church, but the downfall of the Roman order, which was responsible for the backwardness of scholarship and science during the middle ages. In fact it was rather the fall of the Roman Empire and the coming of more primitive peoples which tended to the barbarisation of the Church. Precisely because the world was in an undeveloped state, the herd-spirit prevailed and religion became the religion of the group, nations being converted wholesale on occasion, and Christianity functioning as the cement of society, the buttress of the established order. Precisely because the Church preserved something of the ancient culture in a barbarian world, it enjoyed a genuine intellectual leadership, and there was something to be said for the authoritarian system that it established. One of the results was that civilisation in western Europe developed under the presidency of the Church, and the culture which evolved in the middle ages owed many of its virtues to the fact that it was a Christian culture. It is important to note, however, that this predominance of the Church in a more primitive society where the herd-spirit was so strong, was the result of the special circumstances I have described.

That whole system, which prevailed in Europe in the middle ages, brought many benefits to our world, and in the circumstances of the time it was perhaps the best thing which could possibly have happened for our section of the globe. In a sense the system was hardened rather than mitigated at the Reformation, because religion then came to be organised more definitely on a national basis. Government decided that Holland should be Calvinist, Sweden Lutheran, Austria Catholic, and England under a church of its own. The centuries of modern times are much less attractive than the middle ages in a certain respect, so far as the ecclesiastical side of the story is concerned. Civilisation had reached an adult state; men were differing in their views about Christianity; some men were in revolt against religion. Yet for generation after generation Churches went on fighting to preserve that coercive power, and those properties and privileges which they had held when society was in a more undeveloped state—they went on claiming that the whole of a given nation should be solid and unanimous in religion, allowing no dissent, no departure from the corporate faith. The warfare went on until comparatively recently, and its effects were with us until almost the other day. Not long ago it was necessary to pretend at least that you belonged to the Church of England before you could qualify for certain offices in the kingdom. If you lived in one part of the country, you had to go to the Methodist chapel, because otherwise none of the villagers would come to your shop. People would be induced to conform to religion for all the wrong motives; and it even happened that it required great courage and moral strength to raise one's voice against religion or against the prevailing creed. The protest against all this has coloured our secular literature for two hundred and fifty years, has soaked into the teaching of our curriculum history, and still affects many sides of our education. And the protest has not been frivolous or unscrupulous—often it has itself been possessed of real moral quality. And so today there is often a thick covering of genuine resentment and suspicion which has to be pierced before the Gospel can begin to find a sympathetic hearing.

This is one of the greatest problems that Christians have to face—the resentment and the suspicion of well-meaning men. We can help to solve the problem by improving the kind of history that is current in the world; for it still remains true that the fight for freedom of conscience and the fight for the other liberties was first fought by Christians and was fought on Christian principles, even though the enemy to be chiefly fought was the Church or the alliance of Church and State. We can show the Humanists that all the time the real enemy was the worldliness and cupidity of human beings, and not the spiritual principles of the New Testament. I personally think that we ought to go further and rectify some of our own ways of idealising ecclesiastical history. I think that Christians ought to show the world that they are prepared to renounce more of their history than they are usually prepared to do. We ourselves are very slow to surrender our dream of a Christianity that is allied with power and privilege, though this is the one thing of which there is no sign or trace in the New Testament. Let us stop thinking that religion was in a better state in the old days, when certainly the great numbers came to Church but so often with the result that religion was debased and became less purely spiritual. It is better that Christians should be as they were in New Testament days—humble rather than proud, poor rather than privileged, claiming no rights against society, no rights in the world save that of worshipping the God in Whom they believe and preaching the faith they hold.

The fault of so many Christians in history was that they had not sufficient faith in the power of purely spiritual factors and forces—they wanted to help them out with the strong arm of the law. The same may be the fault of Christians today; yet the long-term results of history seem to me to vindicate the power of the spiritual more than anything else. The real victories of Christianity in history are the quiet ones—the victory of charity, for example, which works like a leaven in society until it leavens the whole lump, or the victory of the Christian martyrs who triumphed by virtue of their very defeat.

There is one further point about the history of Christianity which I think should correct the views of both Churchmen and anti-Churchmen, encouraging the Christian and giving pause to the secular Humanist. Until comparatively recently we studied history from the point of view of government, and indeed largely as the history of government; and even in the case of religion we chiefly studied the government of the Church, the questions of ecclesiastical statesmanship, the policies of Popes and Archbishops. Nowadays we extend our view over the entire landscape—we see the length and breadth of England, the villages and towns, the agrarian units and business firms and centres of local life. And, on the religious side, we see the countless churches that endure for century after century, centres of local worship where the Gospel will have been preached and the Bible read and Christian morality discussed, sometimes week after week for anything up to a thousand years. The real influence of religion lay there, subtly affecting man's view of his destiny, his idea of the good, his notion of how to comport himself in the world, his relations with his fellow-men. Century after century the minds of people have been soaked with the picture of Christian humility, with the view that mercy is greater than judgment, with the conviction that all men are equal in the sight of God, with the thought that they must remember their own sins before condemning other men, and with the idea that love conquers the hardest heart, that love is the only form of righteousness that matters. Just as today we should feel that the policies of kings matter less than we used to think in the rise and fall of states—the economic condition of the whole countryside matters much more—so we would say that the policies of ecclesiastical statesmen are less momentous than the age-long penetration of a whole society with Christian influences springing from every local Church. Indeed it is the Christian principles pervading society which have overthrown the pride and pomp of ecclesiastical authority itself, and which have done so much to shape the ideals of even the secular Humanist. Countless humble men whose names are not remembered in history and who never could have seen the results of their handiwork are responsible for the deeper Christianising of Western society in bygone centuries, responsible for much of the character of even the secular idealism of our time.

16

The Prospect for Christianity

The people who are gravely unhappy about the position of Christianity in the world at the present day are often unaware of the degree to which they are comparing it with the position when the Church had, so to speak, a monopoly through its alliance with power. Certainly the modern world is one in which it is harder to be a Christian than in the days when the herd-spirit and the authority of government and the influence of custom were all on the side of religion. But Christianity offers a higher challenge to the individual and in the new situation of things it confronts each of us with a higher test. The response to Christianity, when it does come, comes with an added authenticity. For anything I know, things may have to become worse for the Church—or rather, I would prefer to say, things may appear to become worse—before we can expect them to be better.

It is obviously of the greatest importance that we should adjust our minds to the kind of world which is now emerging, and not go on thinking in terms of that bygone order under which Christians imagined that they had a right to have everything arranged in their favour. It would be unfitting for Christians even to resent the present régime of liberty, especially as it is one which brings the human race to a subtler state of organisation, subtler because less hard and less authoritarian. Rather we must welcome the present régime of liberty, and say, "This is the Christian thing; this is the result of the operation of Christian principles; the Christians take their stand on 'freedom of conscience' and first made it their watch-word; only the present kind of society does justice to the value which Christianity places on personality, the respect which it insists on giving to personality." In fact the whole transition is one which for Christianity represents a return to the original state of things—a return to something like New Testament conditions when our religion came without any adventitious supports and addressed itself to the consciences of men, calling them to forsake custom and the group, to forsake if necessary even the family itself for the sake of the faith. As in New Testament times, religion is no longer the cement of society, the bond of the tribe, the unifying principle in national life, so that it is thrown back on its essentially spiritual function; it is no longer to be prized for the service it renders to mundane society. Greater responsibility now rests on every single person as he makes his major resolutions about himself, about life in general, about his relations with God, about his rôle in the world. We who are Christians must be the first to say that it is we who insist on this heightened responsibility, this heightened view of personality.

In these days, therefore, we have much to learn from the Christianity of the very earliest centuries; and it would be wiser for us to keep that early period of Church history in our minds as our basis of reference rather than to be hankering after the middle ages or attempting to hold on to an order of things which is now disintegrated. The present age may offer a great test to Christianity but it also offers an opportunity that is possibly greater than any which has hitherto existed. And we do not work without any support from previous history; the Christian Church has already shown that it is able to prevail by an authentic process of individual conversion in a hostile, highly developed, highly urbanised civilisation. Although they were without mundane weapons—perhaps *because* they were without such weapons—the early handful of Christians working in the pagan Roman Empire, and starting from nought, gradually won over a civilised world to the religion of the New Testament. That is the chapter of history that we ought to have before us at the present day, and not the middle ages or the early modern centuries which would be out of context in the twentieth century. I am not clear that there is any argument against the existence of God at the present day which was not known in its fundamental form in the days of the New Testament, and some of the arguments are attractively summarised in a part of the Apocrypha. It is clear that there were certain things in Christianity that contravened the prevailing philosophies of that ancient world in any case; and the New Testament makes no bones about the matter—to the Greeks the whole Gospel was bound to appear as foolishness. There is only the one disadvantage under which we have to labour today and which the early Christians did not suffer from, namely, the history of the Church, the history of fifteen hundred years of ecclesiastical domination, which has on the one hand tended to identify religion too much with conventionalities and with ordinary respectability while on the other hand it has left a legacy of suspicion and resentment. Let us make no mistake about it: in all ages there have been disastrous evils and disadvantages; in all ages there have been forms of worldliness sufficient to daunt anybody attached to the life of the spirit; in all ages the redeeming feature of the story has lain precisely where it lies at the present day—namely, in the fact that amid superstition, indifference, mere conventionality, and worldliness, there always were genuine Christians attached to the spiritual life, genuine Christians doing what they were always told they would have to do, bearing their cross, not knowing whether their work would have results, not always living to see the results, but just leaving the consequences to Providence.

Where Christianity found almost insuperable obstacles in New Testament times was just in the countries where there was a solid religion of state, a corporate national faith; and that was why the Holy Land put up such a resistance—the ancient Judaism was too firmly cemented into the whole structure of the country. The same thing has been true of missionary enterprise throughout the subsequent centuries; Christianity has always found it difficult to penetrate the solid block of Islam. Precisely because civilisation was more highly advanced in much of the Roman Empire, with more scope for individualism in religion, the early Church made its remarkable expansion in Asia Minor and Greece, in Africa, in Italy, Gaul and Spain. Indeed it was the situation of the Roman Empire which was more comparable with the secular world of the present day. Secularism is very hostile to Christianity at the moment, and partly for understandable reasons, but it is fickle and flexible and amorphous, generally unhappy, always flitting like a lost soul in the world, always tragically unsure of itself. It is always hankering to discover a god or a mystique or a form of self-immolation, liable to sink back into dark astrologies, weird theosophies, and bleak superstitions. That no doubt is one of the reasons why Christianity found its opportunity in the ancient Roman Empire. I think that the spread of secularism offers Christianity not only the greatest test but also the greatest opportunity that it has ever had in history. And sooner or later I think it will be realised that the ideals of the noble secular Humanist owe much more to Christianity than is usually recognised, and that they wither—as so often one can see them wither in history—when they are cut off from their source. They wither above all as the world slides blindly into materialism. They are a secularised form of Christianity which proves ephemeral when it is cut away from its original context; and it is the man who I call the "lapsed Chris-

tian" who helps to mark the drift to materialism and the bank-
ruptcy it implies. But if secularism offers to Christianity the kind
of opportunity which the Roman Empire once offered, I am not
sure that the occasion does not require of us a sort of unwinding
of our history—a return to the sources of primitive Christianity. A
return to Biblical and Patristic sources is more relevant to our
time than any clinging to the traditions of the middle ages and
the subsequent centuries.

If we look outside the range of Christendom and extend our
survey to the world at large—the realm of Islam, the awakened
peoples of Asia and Africa and the great section of the globe that
is under communism—we can hardly fail to see that, just as Christianity has lost its privileged position in Europe, so western man
has lost that easy primacy which he once seemed to possess in all
the quarters of the earth. This has happened to a considerable
degree because the other continents have followed the lead of
the West and have advanced so far in the same direction. But they
have followed the West chiefly in its modern mundane lines of
development; our science and our secularism have proved to be
more easily communicable to other continents than either Christianity itself or the subtler virtues of our civilisation. But if in
the last two hundred and fifty years the advance of science and
secularism have undermined the predominance of the Church in
Europe and dissolved that solid block of almost hereditary Christianity in our section of the globe, sooner or later the same
things may well have the same dissolvent effects on the traditional
systems of India, China, and Japan. If science and technology and
the western type of rationalism have so transformed what we
regarded as Christendom, it is difficult to believe that sooner or
later they will not have a similar effect on the other great religions
of the world. In Asia, for example, the very secularism that is
being promoted is a secularism imported from the West and
shaped to the rationality of the Western mind; even an Asia that
is communistic would be an Asia that has given itself over body

and soul to something essentially western. If Asia ever goes western in this kind of way, it will be far more open to Christian
conversion, far more vulnerable to the Christian challenge, than
the strong block of Islam has been during a period of over a
thousand years. In fact the real conflict of religions, as it once
occurred in the Roman Empire, is bound to take place again
sooner or later, but will take place this time on a global scale. The
world may go secular but it will not stay secular. The hungers,
anxieties, and nostalgias which favoured the success of Christianity in the Roman Empire are going to operate in the same way
again over a still wider world. And in the world-conflict of religions which is bound to come and which can hardly be said to
have begun as yet, our Church will have no special privileges. It
will demand only freedom of conscience. In the ancient Roman
Empire it did not need even that, for it won its victories partly
through the readiness of Christians to accept martyrdom.

We do not know how much the traditions of Christianity may
not still be enriched by the many things that we have to learn
from the mysticism of Asia and from the other religions of the
world. If historical Christianity has suffered through being tied
too closely to nation-states, social systems, and prevailing régimes
in Europe, we are now able to see how it may have been constricted through being developed so exclusively in that Graeco-
Roman civilisation which is the basis of the traditional culture of
our own continent. It is perhaps a good thing that in various parts
of the world Christianity is going to be developed by peoples
less subservient to western leadership. It will be interesting to
see what our religion will mean when worked out in the terms of
a culture far removed from the Graeco-Roman. It is even possible
that soon, if not now, the most interesting developments in
Christianity will be those which are not taking place in the West
at all—and this will be the more likely to occur if the West has
become too much bogged in its old traditions.

If Christianity ought not to be tied too closely with the social system, the political régimes, and the nation-states of the West—things which we happen to like because our minds have grown up with them and been shaped to them, so that we can hardly imagine a life that assumes an entirely different form—it is also true that Christianity is not necessarily tied to Western civilisation. Indeed it would be interesting to learn how the first principles of Christianity would have developed if they had fallen into a culture different from the Graeco-Roman. The twentieth-century situation seems to me to require what I have called the "insurgent" type of Christianity—not the kind which binds up its fortunes with the defence of the *status quo*. By "insurgent" Christianity I do not mean noisy or cheap agitation and the pursuit of novelty for its own sake. I mean the kind of Christianity which, instead of merely cherishing tradition and idealising it, is constantly ready to return to first principles, to make a fresh dip into the Gospels and the New Testament revelation. In this sense Roman Catholicism may sometimes represent the insurgent form of Christianity; and English non-conformity may have its moods when it is quite the reverse. But it has always been one of the great features of Christianity that when the world confronts it with a new situation, then, though Christians themselves may greatly dislike the change and try to resist it, the religion can meet the crisis—the Church can go back to the original fountain and source of its life—it can return to the primitive simplicities of the faith and the essentials of the spiritual life, and see how they operate in the new context.

In fact it is not the spiritual life that tends to be conservative. The fault arises from that admixture of worldliness which comes to be mingled in with it almost unaware. The errors of Churchmen in the past were always like the errors of the Old Testament Hebrews—the tendency to interpret the revelation and the promises of God in too mundane terms, so that religion came to be weighted down with too much earthiness. When the Protestants attacked the Copernican theory they were interpreting the revelation in too mundane a way, so that they tied their Christianity to an ancient view of the cosmos, one which in fact was neither Christian nor Jewish but had been inherited from pagan Greece. When the Catholics resisted the rise of modern science they were not basing themselves on anything Christ had said and the scientists were able to quote even St. Augustine against them. They were tying Christianity to a particular ancient theory of the physical universe, the theory of Aristotle, who in fact had not been a Christian at all. Those members of the Church of England who thought that the Bible enjoined monarchical government were making the same mistake—mixing mundane elements into the essentials of Christianity. The progress of modern centuries has been extremely happy in one respect—it has made it impossible to accept the Bible in anything but its spiritual sense. We should not go there now for our natural science, our politics, our economics, our infallible historical narrative. And constantly Christian thought has had to repeat the process of disentangling the essential of the spiritual life from the mundane institutions and intellectual systems—from the earthiness—with which it has become intermixed. Natural science, critical history, academic scholarship, and modern revolution have destroyed everything that Christianity was ever connected with, every superstition that became associated with it—everything except the spiritual life, which emerges better for all the purifications it has suffered. And here we can measure the inadequacy of the liberal theologians who held sway a generation ago. The liberal theologians of the nineteenth and the early twentieth centuries were wrong not because they were intellectually adventurous but because they were not spiritual enough. They did not sufficiently recognise even the data of the spiritual life; they were too much governed by the thought and the ordinary common sense of the world. They would have tied Christianity to the things that happened to be fashionable in the year 1900. But if there is any field that requires daring adventures of the intellect in the twentieth century it is Christianity—not Christianity as a mere organised mundane system, or as a moral programme, but Christianity as a spiritual religion, a contact with "the living Christ."

I have no right to speak of these matters, and indeed I only speak of them with hesitation, but I think I must take it as a basic assumption that, for the Christian, the essential thing is to achieve communion with the living Christ. And the living Christ is as the saints of old described him—continuous with the historical Jesus, yet one with whatever God exists, showing us in this earthly life all that the human mind could grasp of the nature of God. The Christ whom we approach in the twentieth century is not a kind of demiurge, essentially unconnected with the Jesus who walked in this world. And the historical Jesus, though from a mundane point of view he might be in every respect just one of us, still has another dimension and exists in another dimension. He is the living Christ whom we can approach today. It is not an accident that, through tremendous controversies, the early Church sought forms of statement which would safeguard both the historicity of Jesus and the divinity of Christ.

We ought to be clear that we have an important affirmation to make, and that this affirmation is calculated to alter the quality of human life on this earth. And there is a certain sense in which, with that affirmation, we confront modern science, modern technology, and the kind of rationalism that is generally current in the twentieth century. That is to say, while accepting and taking over all these things in their proper sphere, we are ready to challenge the presumptuousness of some of their agents, and to keep these things in their place. But, holding on to this one piece of rope, this one affirmation of ours—taking our stand, so to speak, on this one rock, the living Christ—we are better able to be free and flexible about everything else. We can prevent ourselves from making gods out of mundane things or out of mere abstract nouns. It is important that we should recognise our liberty and exercise it thoroughly. It could be argued that this basic thesis of ours—this affirmation of the living Christ—must govern our minds in certain ways or lead them in certain directions or carry certain corollaries. But even these turn out to be a widening of our liberty and an enlargement of the possibilities of life. If you hold with this teaching you are compelled to move to a higher view of the nature and potentialities of human personality. You are compelled also to have a very spiritual idea of God. In fact it was always true that the things our religion said about the living Christ were impossible to fit into the kind of materialistic and mechanistic picture-making that people generally do when they think about God. An indolent Christian might make a hard and concrete picture of God, but when he thought about the matter he was always driven to subtler and more ethereal realms, where the picture dissolved; God escaped all our definitions and became dematerialised. Our formulas could not hold him and we held our breath before a mystery. And yet, if we talk about the living Christ we do mean that God cannot be in all senses remote and inaccessible, without possible contact with human beings and human history.

What the Christian affirmation really involves, therefore, is the insistence that life in the world is capable of a further dimension which many people leave out of account. We all understand that man has a life that can be lived on an intellectual plane, transcending mere animal existence, though I suspect that if we had looked at this globe some millions of years ago none of us would have believed that even that was possible. None of us would have believed that lumps of matter could ever have stood up in the world (as we do) and asked what it was all about, ask what they were doing there. We all understand that man has a status not only as an intellectual but also as a moral creature, though this would not necessarily have anything to do with religion, since morality itself is often held to possess merely a mundane reference. The Christian challenge, however, has meant the assertion of a further dimension to the universe and to human

life itself—a spiritual order of existence in which men have felt
that they walked with God and had communion with the living
Christ. Religion may need continual reformulation and restate-
ment, but we are not permitted to evaporate that spiritual element
out of it. And the ultimate intellectual problem is the question of
the validity of spiritual experience. Churches are sometimes dangerous things, and, unless they are very spiritual indeed they can
actually be a harmful influence in a country. But where they are
unique, where they really stand or fall, is in their assertion of an
additional dimension to life—this communion with the living
Christ, this insistence on the actual practise of the spiritual life. I
am so sure of this—and yet so dubious about everything else, so
dubious even about everything else in the Christian tradition—that
if I had my way we should say to young people at the present
day: "Lay hold or just this one thing, and carry it where you like.
Try to grasp this spiritual dimension of life, throw everything
else overboard, and see where this takes you."

Some people see a mystical union in the Church of a kind which
enables some men to have power over others, or strengthens the
domination of the past over the present—they are prepared to talk
about the submergence of the individual in the mass. Nobody could
be more suspicious of such thinking than I; and often I have felt
that it led people to put conditions on salvation through Jesus
Christ, or on fellowship with Christ—conditions of the very kind
which our Lord set out to overthrow. But there *is* a mystical union
in Christ, and it is a thing which is realised in actual Christian experience. And it is a union that we can find not only with our contemporaries but also with Christians in all the centuries. It is not to
be confused with the herd-spirit which is found at a lower level of
existence and which in primitive societies does mean that the
individual is submerged in the mass. It is a union entirely different
because it is based on love, like the union that can exist at a different level in a voluntary and effortless way between a couple
of sweethearts. It is a union of free personalities bound together

in a love which, far from submerging the individual, carries personality to a still higher power. Arrogance of intellect is inconsistent with Christianity, and the Christian will certainly stand in some sort of humility before the faith of the saints and before the consensus of Christian witness. If he differs from these he will do it only after prayer and fasting, and as one who takes his soul in his hands. But it is also his duty to be quite daring as a Christian thinker, using his mind as well as everything else in the service of Christ. I do not myself conceive that within any foreseeable time there will be amongst the most genuine Christians a real unanimity even on what Churches regard as essential statements of doctrine. The union of Christians is a union in love and in the spirit but not in intellectual formulations, and I personally hope that no set of men will ever make surrenders on points of important doctrine as though these were trifles—indeed such surrenders would only tend to precipitate the foundings of new sects again. There is a union amongst Christians which overrides what I must call intellectual differences, overrides the various rationalisations that men try to make of their faith. And that spiritual union is the fundamental thing. We all know that our rationalisations and formulations of it are bound to be imperfect in any case. Where the spiritual union does not exist it is pointless to speak about acquiring it by force; but it is even vain to think of acquiring it by a lot of argument or argumentativeness.

In conclusion, therefore, today, if Christianity in our part of the world has lost ground in point of numbers, it has shed at the same time many of the evils that went with numbers—superstitions, attachment to vested interests, and a great deal of that admixture of earthiness. The saddest feature that I have seen at the present day has been the discouragement of some Christian workers who carry the real burden of the transition and are not always in a position to see its better side, so that sometimes they even come to feel that they are working to no purpose. I wish I could convince them that this is the greatest moment in a thousand years for the preaching of Christianity and they can safely leave the results to Providence. For a thousand years the Church was able to rely on custom, on the prevalence of religious routine, and on the persistence of what I might call the conventional Christian. It is the conventional Christian that has largely let the Church down—he has been leaving the fold—and convention now rather seems to be the thing which pulls people the other way. It is possible that the organisation of the Church, and its techniques and policies are directed to cater rather to the man who is Christian by inheritance, so to speak, and by convention; and sometimes I have wondered whether the pronouncements of various Churches would not have been rather different but for the fear of offending the conventional Christian. Nobody counts the number of noble and quixotic intelligent young people who are put off by this—there is no means of counting them because they disappear in a sort of conjuror's hat. I sometimes wish they would be brave and come in and take charge of us—come in and just grasp the helm themselves. At any rate it is worth while noting that in the present state of the world it would be a tragic thing to be guided more by the fear of losing present church-members than by the confidence that there are new ones waiting to be won. There is a Providence in the historical process which makes it more profitable to be guided by one's faith and one's hope than by one's fears.

17

Christians in the Coming Period of History

The time is coming—and in certain respects is already here—when we must expect it to be much harder to be a Christian, much harder than in former times at least to make the initial commitment to Christian belief.

This is a fair thing, a fair challenge for Christianity to have to face; and it might be argued that in the past what was really too paradoxical was to see the faith of the New Testament backed, for example, by the power of the state, or by the force of social custom—people even drifting into Christianity sometimes because that was just the respectable thing to do.

In days gone by, you might be born into Christianity and brought up without any notion that there could be any reputable alternative to it. Under these conditions you might even slide thoughtlessly into an inherited creed, accepting it as one of the conventionalities, or tending by your own conduct to turn it into something rather conventional.

In the future Christianity will be one of a host of things that present themselves to human choice, and will have no mundane force to rely on save the power of its own message. It will address itself to people who are postured for asking the question, trained to ask the question: Why should we believe this particular claim to revelation which comes down to us from ancient Palestine? They will in any case ask the further question: What has Christianity got to contribute to the late twentieth century—what has it to contribute of its very own, apart from the ideas that Christians themselves simply take over from the non-believing world? It is going to be a great testing-time for our religion though it might also be a creative period. Possibly we are returning to something more like the situation of that first handful of Christians who carried the Gospel to the brilliant highly civilised parts of the pagan Roman Empire. The position may even be more difficult still; for, in the first place, Christianity now has a long history and some people will take a long time to forget or forgive the power that was long held by churches and some of the mistakes they made in their use of that power. In the second place we have to face the physical universe of modern science, and also the criticism to which scholarship has submitted our own basic sources—history presenting us with problems perhaps as imposing as those of modern science. In the future, then, it is not going to be so easy to be a Christian save by a high and autonomous act of decision—a decision that a man may have to make in his solitariness, after grappling with all the issues, after really confronting himself, confronting the universe, confronting God. There will be no easy way of sliding into Christianity while avoiding the confrontation.

All this does not involve us in anything which in itself is radically new, however, and there is a sense in which a drastic confrontation has always been necessary for anybody who tried to bring his Christianity really home to himself. It has been the great theme of religion that the heart of the world's difficulties lies in the problem of human nature, average human nature, universal human nature. Just the ordinary cupidities of men—the mere desire to better one's own position a little, for example—even only the fear of having to reduce one's standard of living—this kind of thing, multiplied by millions amongst the inhabitants of a nation, can pile up into a colossal pressure on government, driving a state *even* to aggressive policies. The same general defects of human nature saddle us with tremendous problems of human relations, and these follow very much the same fundamental pattern whether in ancient China or in modern Europe—they arise within societies of wealthy and educated peoples or even in some local church, as well as amongst the victims of an industrial system— they are to be found in democracies as well as in monarchies. In other words the trouble is not eliminated by mechanical changes in the circumstances under which men live. On this whole general issue, ancient Palestine did make a historical contribution: and the Bible, first of all in the Old Testament as it proceeds, and still more in the New, gives a surprising turn to the argument. While always unrelenting about the character of sin itself, the Bible becomes kinder to the sinner as time goes on and as it reaches its peak places. And, already, within the limits of the Old Testament, the picture of God as the terrible Judge is being superseded by the picture of a God who pulls on human beings like a magnet, drawing them with the cords of love. It was quite a remarkable thing that Jesus put his finger on self-righteousness as a crucial evil—pin-pointing the special danger of it in religious people and apparently amongst a class of them who were otherwise rather noble characters.

Therefore, it has always been fundamental to the Christian religion that no man can begin to approach this problem of human nature unless he has first of all tackled it inside himself, recognising that he himself is part of the problem—he has his share in man's universal sin. If anybody is bent on either judging other people or improving the human condition generally, he has a particular call to look at his own personality in the first place and to take great care of his own interior—otherwise his internal compass is liable to be deranged from the very beginning, deranged, for example, by concealed egotisms and facile self-deceptions. Some people, for example, are governed more by their hatreds than by their loves—there are some who hate the capitalists more than they love the poor. And perhaps, too, we do not sufficiently sting people with the real paradoxes of Christianity—the fact that a man can give all his money to the poor and still not have charity, not have love, in the Pauline sense. And it is not merely a question of rigorously analysing one's own motives and cleaning them up. There is the whole business of getting below those successive layers of insincerity with which we so constantly deceive ourselves, a case of getting behind the screens, the masks, the camouflage, that we set up in our own minds—the tricks of conventional language, the current slogans, the unconscious posing and play-acting that we do. It seems to need a lot of trouble, to take a lot of penetration, to see ourselves as we really are—indeed to see anything naked. And that is why there has always been an element of spiritual self-cultivation in the tradition of Christianity, not for any self-regarding purpose, but to make sure that one started with one's own personality properly balanced, that one found one's own centre before pretending to give direction to others, that one had one's own internal compass properly set, so that one confronted the problems of the world with adequate spiritual resources and internal strength.

Even in traditional Christianity, then, there was often the call that brought one to the awfulness of direct confrontation—confrontation with oneself and with everything that is not oneself— the kind of thing to which men are in any case driven sometimes by a colossal national tragedy that leaves no room for any further self-hoaxing. It is true that some people can go through a whole lifetime without having had a direct confrontation with anything. Those who live in the 1970's, like those who lived in the year 1700, may just slide into the easy philosophies of contemporary journalism, picking up ideas just because they happen to be in the air. But in the coming period Christians themselves are going to be subject to pressures that must send them back to their original sources and their first principles. They are going to be forced to dig very deep before they can turn the present crisis into a creative moment. I wondered whether, for that particular kind of encounter, there were not some fairly elementary points—essentially perhaps points of method—that one might do well to keep in mind.

First of all, I think we ought to make enquiry about everything possible, pile up our knowledge in one field and another, and, if you like, become as sophisticated as we can in the realms of science and learning. But, wherever we may stand in respect of all those things at any given moment, it is still true that when we come to making ultimate decisions, our eyes ought to pierce through all that intellectual luggage as though it were sheer transparency. The information ought to have churned itself simply into experiences, losing itself because it has passed so completely into the very walls of the brain. And then, holding our breath in the quietness we ought to recapture the humility, the simplicity, and the directness of a little child—the child in the story who just said that the king had not got a shirt on, the child that Christ said we had to become before we could enter the Kingdom of God. We ought always to remember that the greatest single cause of intellectual aberration is any sort, any trace, of intellectual arrogance. If this seems too juvenile a thing to put forward, I would add in the first place that it requires a high degree of what you might call artistry to become as a little child. Furthermore, I am not sure that some of our greatest scientists would not say that they reached their most original ideas rather in the way that I have just described.

Secondly, in any case, no views about the ultimate things of life are of any use unless they are formed not merely in an attitude of intellectual humility but also in a mood of actual love towards the universe, including affection for human beings in spite of their faults, including also a certain wonder and delight at the fact that, out of the very dust of the earth, here we are, lumps of living self-consciousness—here we are, bits of this old universe,

standing on our hind legs and asking the universe what on earth
we are doing here. This self-consciousness of ours, this self-
consciousness, rather, which each of us possesses in his solitariness,
as though it were a separate island, closed to everybody else—this
self-consciousness which can not be reduced to anything else—is
the most interesting piece of information that we possess about the
whole order of existence. In any case, the slightest hint of resentment against the universe in one's heart—the slightest trace of anger
at having been born into such an imperfect world, or even indignation against human beings for not being better than they are—anything of all this would always be a further cause of intellectual
aberration.

Thirdly, no view of life and the world is worth very much
that does not do justice to the ultimate mystery of things. Sometimes I get the feeling that in every generation the scientist thinks
that he has found the essential clue, that he is close to the final secrets. The picture only needs filling in or rounding off. And you
certainly find that men erect their over-all philosophies on the
basis of the state that science happens to be in at just that moment. There is an almost ineradicable tendency to see the historical process as manufacturing a future that shall be like the present, only more so, everything going in the same direction as now,
but far more quickly. But also *Christians* have often badly contravened their own rules—imagining that they have got everything
taped, that they know how God does his thinking—so that they
sometimes drive the mystery out of religion itself.

Fourthly, if you look at the human race over a long stretch of
time you will find that for one very considerable period, say
from 600 B.C. to 300 A.D., particularly in Asia, but towards the
end of this period in Europe too, men had more anxiety about
their destiny and about the meaning of life than people today
would have or would imagine to be possible. And they set out to
explore the spiritual side of life, the caverns and corridors inside
their own personalities—set themselves to feel out and find God—

doing this with all the intensity, all the single-mindedness, all the genius with which our generation explores the purely physical side of the universe, the scientific side. I think it is a pity that the human race so often imagines that it has to concentrate on just one aspect of things at a time, and this is one of the mistakes that helps to make the world so unbalanced. But if we do happen to live at the scientific stage in the history of the human race, this is no reason why we should throw overboard the things which once occupied intellects as great as any that exist at the present day. It is no reason why we should reject altogether a vast area of human experience, and a thousand profound insights that were gathered at an earlier stage in the story when people were equally intent upon other aspects of experience.

Fifthly, I would personally be affected by a thing which other people have felt but which has elements in it that are always almost incommunicable. I happen to have known a number of people—two of them amongst the ablest men I ever knew, but two of them who never had any schooling after the age of ten, one of them a woman who never read anything apart from the Bible, except a very popular paper called the *Sunday Companion*—and they seem to me to have reached the real secret of life and touched the bottom of things in a way that makes me feel the world of forlorn clever rudderless dons quite a superficial affair. I could not convince anybody else but I am convinced myself that these people walked with God, and there are a lot of them in history—these people who were sure of their contact with the living Christ. The historian has to admit the extraordinary power that this had over their personalities and lives. What would be questioned by historians who are non-believers is only the way these people diagnosed themselves, the way they explained the power that was within them. In this sense, the crucial question is the question of the validity of religious experience.

We take our stand—we cannot have a tenable doctrine of revelation unless we take our stand—on this further dimension that life possesses here and now: God letting men make their own history, make their own messes in history, letting them face all the risks and chances of physical existence in quite a dangerous universe, but pulling on them like a magnet, and speaking to them, speaking only to the innermost parts of them. Apart from that, we do not expect him to interfere with the realm of ordinary causation in order to give the twentieth century the general turn that we ourselves would like it to take. But in respect of his presence in the devotional life of the Christian, there is no difference between one century and another.

It is not clear that we can actually assist religion by basing it upon the utilities, which sometimes turn it into banality or reduce us to a kind of baby-talk. History would have a lot to say in support of the case that a supernatural religion is quite a dangerous thing to have in the world—very dangerous indeed, unless soaked and deluged by the principle of charity. Possibly the first significant judgment a man makes about the universe is a judgment he passes on something that he really can know something about: the judgment he passes on himself—whether he feels himself to be just sovereign intellect or recognises that even in his intellectual operations he shares the infirmities of human nature generally. But what we really face the world with, as Christians, is the affirmation of the spiritual life in Christ and if we are open to reproach it would be for our lack of intensity on this side. If I desired to say perhaps one thing that might be remembered for a while, I would say that sometimes I wonder at dead of night whether, during the next fifty years, Protestantism may not be at a disadvantage because a few centuries ago, it decided to get rid of monks. Since it followed that policy, a greater responsibility falls on us to give something of ourselves to contemplation and silence, and listening to the still small voice.

Index

2. Main Ideas and Themes